Low Carb High Fat Cooking for Healthy Aging

Annika Dahlqvist & Birgitta Höglund

Low Carb High Fat Cooking for Healthy Aging

70 Easy and Delicious Recipes to Promote Vitality and Longevity

Skyhorse Publishing

Skyhorse Publishing books may be purchased in bulk at special discounts for sales promotion, corporate gifts, fund-raising, or educational purposes. Special editions can also be created to specifications. For details, contact the Special Sales Department, Skyhorse Publishing, 307 West 36th Street, 11th Floor, New York, NY 10018 or info@skyhorsepublishing.com.

Skyhorse® and Skyhorse Publishing® are registered trademarks of Skyhorse Publishing, Inc.®, a Delaware corporation.

Visit our website at www.skyhorsepublishing.com.

10 9 8 7 6 5 4 3 2 1

Library of Congress Cataloging-in-Publication Data is available on file.

Cover photo credit: Mikael Eriksson
Interior design and photos: Mikael Eriksson, M Industries

Print ISBN: 978-1-63220-533-9
Ebook ISBN: 978-1-63220-934-4

Printed in China

Annika Dahlqvist, MD, on LCHF

How LCHF Became a Concept

I am a general practitioner who was once overweight. I used to experience many aches and pains due to IBS (Irritable Bowel Syndrome), gastritis, esophagitis, fibromyalgia, fatigue-induced depression, sleep disturbance, and an irritable bladder. All this negatively affected my quality of life. I was eating the way we were advised to: low fat, margarine, and vegetable oil, as well as lots of whole grains, vegetables, and fruit, yet I kept getting sicker and sicker.

In the fall of 2004, my daughter, who was a medical student, advised me to try a diet that included very few carbohydrates and didn't restrict the consumption of saturated fats. My life was transformed: after two months, I had not only lost weight; all my aches and pains were gone. This sparked my interest in nutrition; I met Uffe Ravnskov, MD, PhD, an independent researcher who told me that, contrary to opinion of most "nutrition experts," natural saturated fat is not bad for the heart at all. In fact, it acts more like a protector of heart health.

After studying nutrition and gaining knowledge and experience in the field, I began to share my information through a blog, as well as in lectures and books.

LCHF stands for Low Carbohydrate High Fat, and describes a diet that contains few carbohydrates and more fat. It's a model for eating that has been developed in Sweden since 2006 when I, along with other dedicated parties I met online, launched the idea.

The following is information about the meaning of the LCHF concept, and why it's so healthy for you. You'll find many excellent recipes for everyday meals, and old favorites, too. Many happen to be my personal favorites. You should have no problem finding the ingredients for these recipes at your local grocery store.

Good luck and bon appétit!

LCHF is Beneficial for Most of Us

Anyone who is ill or overweight can benefit from eating according to LCHF (i.e., a diet low in carbohydrates and high in fat). This is sound nutritional advice even if you are not sick, to prevent illness from developing and weight gain from creeping in. Most of the LCHF adherents profess feeling very well; your body will thank you for giving it the food it was meant to have.

The three building blocks of LCHF:
• Fewer carbohydrates
• More natural fats
• Foods free of poisons and chemicals

1. Fewer Carbohydrates

Carbohydrates comprise sugar, starch, and cellulose. Starch is a polysaccharide, a part of glucose that is present in potatoes, grains, corn, and rice. Cellulose (fiber) cannot be digested by the human body, and thus goes straight through without turning into glucose. Sugar and starch can be digested, and they spike our levels of blood sugar.

Foods low in carbohydrates do not elevate blood sugar levels, and by eating them, you'll avoid the insulin response set in motion by high blood sugar. Insulin rapidly lowers blood sugar, which creates feelings of hunger, even shortly after a meal. Insulin also facilitates the storage of fat and blocks our ability to burn fat. Insulin transforms excess of carbohydrates into fat tissue.

Elevated levels of blood sugar and insulin also bring on inflammation, which is a precursor to many of our most common modern-day illnesses: cardiovascular disease, cancer, rheumatism, inflammation in the digestive system, and many others.

The body is fully capable of producing the blood sugar it needs from protein and fats in food. This is called gluconeogenesis. This blood sugar is stable, instead of being subjected to abrupt swings caused by carbohydrates.

Thus, low-carbohydrate nutrition is a perfect fit for diabetics, since blood sugar is not elevated to any appreciable degree after a meal low in carbohydrates.

If you follow the LCHF model, 1 3/4 ounces (50 grams) of carbohydrates is your limit for the day. If you are very sensitive to carbohydrates (i.e., you are diabetic or pre-diabetic), you need to stay below 1 ounce (20 grams) per day.

2. More Natural Fats

When we lower our carbohydrate intake, we need to eat something else to feel satiated. The best way is to replace the calories we got from carbohydrates with fat. Fat is needed to support body growth as well as all the body's functions. Fat doesn't elevate the blood sugar, which makes it an excellent source of energy.

There are, however, good and bad fats, and we need the good ones for our health. There are two polyunsaturated fatty acids that we need to obtain through our food—omega-6 and omega-3. We often get too much omega-6 compared to omega-3. Ideally, the ratio should be 1:1, an equal amount of each. A lot of omega-3 is found in fatty fish and fish oil, meat and milk from grass-fed animals, and in eggs from hens eating omega-3 enriched feed. It's a good idea to take fish oil supplements if you don't eat fish, or eat very little fish. Eating omega-3 improves the immune system. On the other hand, if we ingest too much omega-6, we risk inflammation in the body, which in turn can lead to blood clots and cardiovascular disease later in life.

Butter is good and natural, and is produced in the Nordic countries. It is a pure animal fat that also contains a good amount of omega-3, especially if it comes from free-range cows that are fed grass and hay. Butter is not adulterated or processed in chemical facilities.

Colza, canola, and olive oils are other good fats that are not unhealthy to us. There is a drawback to colza oil, however, in that it contains a toxin that cannot be entirely eliminated, so the effects of consuming colza over a long period of time are unknown as of yet. As for olive oil, its main disadvantage is that it's not manufactured locally, and thus has to be imported from the Mediterranean. Both colza and olive oil should be cold pressed and organic.

Studies have shown that coconut oil can improve memory and cognitive function, and can even stop or slow down the progression of diseases such as Alzheimer's. It should preferably be organic coconut fat from the health-food store. The mass-produced coconut fat for sale in regular grocery stores is not deemed to possess the same beneficial properties.

Corn, sunflower, thistle, peanut, and soy oils contain too much omega-6. Palm oil is extracted from fast-growing palm trees, which are planted where the rainforest has been decimated to make yet more room for palm trees. Margarine, under its many different guises, is an industrial product that is manufactured from palm oil and makes an inferior replacement for butter. Comparative studies show that children tend to develop more allergies from consuming margarine than from butter.

There is no reason at all to believe that we have to eat low-fat food to stay healthy. The American scientist Ancel Keys is the father of the fat phobia that emerged in the United States of the 1950s. Backed up by faulty statistics, he managed to convince nutritionists all over the world that naturally occurring saturated fat was bad for our hearts. This myth has now become today's most dangerous old wives' tale, as it scares people away from eating healthy, natural fats.

Cholesterol is a highly specialized molecule that's necessary for sustaining life in our bodies. The notion that cholesterol is dangerous is another harmful misconception that also came out of the 1950s. It wasn't until in the 1980s that researchers were able to produce a study that showed that coronary attack patients could perhaps reap a tiny benefit from medication with statins to lower their cholesterol levels. A long-range study (The Karlshamn Study) shows that a low-carb diet lowers the risk of coronary events in type 2 diabetics far more efficiently than statins.

Statins have many side effects, the most common of which are muscular pain and weakness. Lower levels of cholesterol can impair cognitive function, since cholesterol is essential for healthy brain activity.

Protein consists of amino acids, ten of which are essential; the body can't make these amino acids and so we have to get them through our intake of food. The perfect combination of amino acids is found in animal protein, which is why children, the elderly, the sick, and the overweight should eat fish, meat, and eggs to the get best quality nutrition possible.

3. Minimize Additives, Eat Organic and Local Products

Chemical additives for use in food have been around since the beginning of food processing. These include taste and aroma enhancers, preservatives, sweeteners, gelling agents, etc.

We don't know the full scope of the effects those additives have on our health. Tests are performed on rats, and if the rats don't die, it is generally taken as a sign that the chemical being tested is harmless for humans to eat in large amounts, each and every day. So as long as there are no long-range studies of the effects of food additives on humans, it's better to avoid them altogether whenever possible.

The easiest and best way to do this is to prepare your food from scratch with clean, organic ingredients.

MSG (monosodium glutamate) is the most widely known and controversial additive. MSG is used to improve the taste of inferior foods. MSG has proven to have many detrimental side effects and is, among other things, highly allergenic. MSG also fools the brain into thinking we are hungry or feeling peckish. There have also been reports of behavior disorders such as aggression, having difficulty in concentrating, hyperactivity, and all the problems that can be found along the ADHD (attention-deficit/hyperactivity disorder) spectrum. MSG can often be found in spice mixes and ready-to-eat dishes, so read food labels carefully. Restaurant meals are also often seasoned with MSG.

Season your food with salt and pepper and fresh herbs. If you start off using quality produce, you won't need to add a lot of extra flavorings. We're starting to see more and more spice mixes without MSG in stores.

There is more to read (in Swedish) on the subject of additives in our foods by the Swedish authors Mats-Eric Nilsson and Tommy Svensson (Utbildningscenter.se).

Vitamins and Minerals

There are more vitamins and minerals in meat, fish, and eggs than in fruits and vegetables, with the exception of vitamin C. Vitamin C is, among other things, an antioxidant that helps us fight the effects of inferior food and free radicals. The amount of vitamin C we require can be supplied by the vegetables we eat at dinner.

Fruit doesn't provide anything important that we can't already get from vegetables. Fruit is loaded with sugar, however, and half of it is in the form of fructose. Health-wise, fructose is the worst sugar you can eat, so avoid fruit if you're sugar sensitive, or enjoy one piece as a treat on a special occasion.

Fruit is also often heavily treated with pesticides. If you eat fruit, make sure it's organic.

Sweeteners

Artificial sweeteners such as aspartame are chemicals produced by food manufacturers. They have both known and as yet unknown detrimental side effects. The two sweeteners that, to my knowledge, have no warnings against them are stevia and erythritol (combined in Truvia sweetener), so feel free to use them. You'll find them in most health-food stores and online.

Nutritional Advice

I like to eat two to three times a day, but there are followers of LCHF who eat fewer times than that, and also those who prefer to eat more frequently by including a few snacks. Try different ways until you find a level that fits your temperament, weight, and hunger level. Many feel satisfied for longer periods of time by following this kind of nutrition. Give your body a rest from the digestive process during the night, but if you're working night shift and get hungry, stay away from carbohydrates and eat some protein and fat instead.

Eat and Enjoy!
You can eat freely of the following (preferably organic and full-fat):
- Dairy products, especially full-fat items such as cheese and butter. However, milk, cultured buttermilk, and yogurt in large quantities are too rich in sugars for anyone who is sugar sensitive.
- Meat from beef, pork, lamb, poultry, fish, and shellfish. Don't cut off any marbling or visible fat!
- Salt, pepper, and fresh herbs and spices
- Homemade sauces without carbohydrates
- Vegetables and root vegetables with low carbohydrate content
- Salad dressings made from mild olive oil and cold-pressed olive oil, vinegar, or lemon
- Olives and avocados
- Coconut fat (from health-food/online stores; not from the grocery store, as it is most likely industrially manufactured)
- Eggs, preferably organic

Eat in Limited Quantities Only

Enjoy a little of these:

- Beans, lentils, and fruit (no dried fruit)
- Nuts, almonds, and sunflower seeds
- Chocolate with high cacao content (at least 70 percent)
- Sausages and deli meats—check the quality and avoid products with added MSG (in EU countries, E #s 620-661, especially 621)

If you want to lose weight, avoid fruit and nuts, as they contain too many carbohydrates.

Don'ts

Avoid or eat the smallest amount possible of:

Foods containing a high amount of carbohydrate, i.e. sugar and starches:

- Potatoes and dishes based on potatoes
- Rice and rice products
- Corn and corn products
- Grains: pasta, bread, crackers, breakfast cereals, porridge, pancakes, etc.
- Sugar
- Foods containing hidden sugars, such as Swedish whey cheese and blood pudding
- Sweets, bakery confections, cakes
- Sugar-sweetened soft drinks

Bad fats:

- Margarine
- Oils with too much omega-6 (such as corn oil, sunflower oil, thistle oil, and peanut oil)
- Prepackaged and prepared foods—keep a close eye on what you eat in restaurants

Health Gains

Diabetes and Metabolic Syndrome

Since you eat so few carbohydrates when following an LCHF plan, your blood sugar will not be elevated, which makes it the very best nutrition for diabetics.

The pancreas of someone with type 1 diabetes is unable to produce insulin, and so that person needs to inject it. There isn't the same symbiosis between glucagon (the hormone that elevates blood sugar) and insulin as when insulin comes naturally from the body. This means that there will still be swings in blood sugar levels even if the person is fastidious about monitoring his or her food and insulin dosage.

Metabolic syndrome manifests itself in overweight—abdominal obesity especially—high blood pressure, and insulin resistance. Insulin resistance indicates a decreased ability to naturally lower one's blood sugar after a meal rich in carbohydrates.

People with type 2 diabetes have such strong insulin resistance that they cannot manage to keep their blood sugar at a stable level—at least not while eating a diet full of carbohydrates. By eating an LCHF diet, most type 2 diabetics are able to manage their blood sugar with as few medicines as possible.

Metabolic syndrome and type 2 diabetes are often precursors to complications such as myocardial infarction, arrhythmia, and stroke.

The best way to eat for someone who is an insulin-dependent diabetic is to lower the intake of carbohydrates to less than what is necessary for normal blood sugar control. That way, the finely tuned level of blood sugar is balanced by the glucagon effect on glucogenesis. A low intake of carbohydrates provides a stable, normal blood sugar level using the least amount of insulin possible. For further information, read *The Diabetes Solution* by Richard Bernstein, MD. As a type 1 diabetic and physician, he discovered that he could be a healthy diabetic by eating a low-carbohydrate diet.

All diabetics can benefit from eating a LCHF diet. Many signs indicate that vascular diseases that often follow a diagnosis of diabetes, such as eye problems, damaged kidneys, and vascular and nerve damage in the legs and feet, can be bypassed by following a low-carbohydrate diet.

Many diabetics can decrease their medications significantly—some may even be able to stop them altogether—after switching to an LCHF regimen. If you need insulin injections, you'll need to decrease the amount of insulin you take before you start eating according to LCHF rules. If you don't, you'll run the risk of lowering your blood sugar too much and becoming hypoglycemic. The insulin decrease needs to be done under medical supervision. If your nurse or physician is skeptical about LCHF, seek out a medical team that is supportive, or else this might turn into a challenge. There are examples of type 2 diabetics who have been able to stop their medication after only a few days of eating LCHF.

Inflammatory conditions such as rheumatism, asthma, and psoriasis can often be improved by switching to a LCHF diet, because elevated blood sugar and insulin cause inflammation and aggravate rheumatism. Different food additives can also bring on inflammation.

Thin and/or sensitive skin becomes stronger and more elastic due to LCHF, because it is better nourished. Leg and foot ulcers heal faster due to improved circulation and nutrition provided by LCHF.

Diabetes is the most common reason for renal failure. Healthier blood sugar levels and less insulin also means better kidney health.

Digestive problems usually clear up on an LCHF diet. Enteritis and IBS (Irritable Bowel Syndrome) improve significantly when the intestines don't have to fight against loads of carbohydrates and fiber. Many have told of their serious and long-lasting digestive issues going away just a few days after changing their daily diets to LCHF. Even conditions such as gastritis and esophagitis can be dealt with successfully, as they are caused by the inflammation and digestive malfunction of a system that is misguidedly fed a low-fat diet rich in carbohydrates and fiber.

Migraine sufferers can also get better by switching to LCHF. At my clinic, the nurse who specializes in diabetic care discovered that she experienced no migraines once she minimized her intake of carbohydrates.

Dental health is also vastly improved by a drastic reduction of carbohydrates; cavities, gingivitis, and periodontitis stop. Starch becomes sugar as soon as it enters the oral cavity due to salivary amylase enzyme action; in turn, sugar, together with the bacteria already present in the mouth, produces an acid that erodes the teeth's enamel and inflames the oral mucous membrane.

Infections. It is an undisputed fact that our bodies' cell membranes are best built from natural fat and protein. A healthy cell is the best defense we have against invading microorganisms—it offers resistance against viruses, bacteria, and fungi. Bacteria and fungi use sugar for energy, but cannot make use of fat. Our whole immune system is built from fat, cholesterol, and protein.

Obesity. An excess of carbohydrates produces elevated blood sugar. Insulin stores this excess sugar as fat in our adipose tissue. By minimizing our intake of carbohydrates, our insulin level is lowered, fat can be used for energy, and our weight is thusly reduced. LCHF is the only way to lose weight without feeling pangs of hunger.

Osteoporosis. Our bone cells are built from natural fats, cholesterol, and protein, but calcium and vitamin D are also required. The best source of vitamin D is from exposure to the sun when it's at its zenith. But many older people don't get enough sunshine, and consequently suffer from osteoporosis. Another way of effectively absorbing much-needed vitamin D is taking it in the form of dietary supplements.

Urinary tract problems. An overactive bladder can be settled by switching to an inflammation-reducing LCHF diet. Even stress-induced incontinence (caused by sneezing, laughing, or heavy lifting) can be dealt with by LCHF, because the diet improves the pelvic small-muscle function.

The same applies to tendencies of uterine prolapse (i.e., the uterus descends into the vagina). The uterus can be anchored in its correct position when the pelvic muscles get proper nourishment from LCHF. However, if the uterus has already descended, a change in diet will not fix this—a uterine ring or surgery will be necessary.

Many different factors can cause *cancer.* Elevated blood sugar, elevated insulin levels, and industrial chemicals are some of the causes. Whatever the cause, cancer cells can rely only on sugar for energy. They cannot use fat. Cancer cell growth is therefore hampered by a low-carbohydrate diet, and stimulated by a high carbohydrate intake. Cancer cells exhibit a preference for fructose, which is found in fruit and sugar. Elevated insulin levels in the body stimulate cancer growth.

Dementia. The most common type of dementia is Alzheimer's disease. It is recognized by the buildup of amyloid protein in the brain cells, which are then destroyed. Amyloid has one important building block—insulin—and excess levels of insulin hinder the removal of amyloids from the brain. There is a clear indication that the risk of developing dementia is multiplied many times over for diabetics who have elevated blood sugar. On the other hand, several reports have shown that coconut fat improves cognitive function in both healthy individuals and persons already suffering from dementia.

Frequently Asked Questions

Don't you get gallbladder problems if you eat too much fat?
Gallstones are formed when you eat too little dietary fat. The gallbladder's primary function is to emulsify fat, so when there is little or no fat to deal

with, bile remains underactive in the gallbladder. Cholesterol forms sediments that crystallize into stones, which are "sanded down" by the movement of the gallbladder. When you eat some fat, the gallbladder discharges bile into the bile ducts and out into the small intestine. During this process, a stone can get lodged in the biliary duct, and the pain you feel is from the buildup of bile behind the stone. Once the attack is over, the pain goes away because the stone has been expelled into the intestine.

When you consume fat, there is more bile activity. Bile flows over the stones and can dissolve them when they're still in the gallbladder, or they exit via the biliary duct—with or without a gallbladder attack.

Many people have reported having gallbladder stones, but after eating LCHF for a stretch of time, a follow-up ultrasound indicated that the stones were no longer there.

But for someone who has many and/or large stones to deal with and inflammation of the gallbladder, surgery to remove the gallbladder or the stones might be necessary, after which they can resume their LCHF way of eating. There will still be the same amount of bile sent to the intestine to work on the fat. If you still feel some discomfort after eating a fatty meal, it might be easier to try eating in smaller amounts over several meals.

Why don't I lose any weight?
Most people have no trouble losing weight once they change to LCHF. However, a small percentage of people still experience some difficulty in this endeavor. Perhaps their bodies simply don't want to lose weight, or there might be something in or about their bodies that prevents them from shedding excess pounds.

My advice is to reduce carbohydrates to minimize the fat-burn-blocking insulin response. Then, experiment to find the right amount of food to eat that's conducive to weight loss for you. The body won't burn stored fat if it gets enough of it to meet its daily requirements, but do not decrease your intake of food to starvation levels—this is very detrimental to your health and well-being.

I can't drink milk, so what can I drink instead?
Being lactose intolerant means that you're unable to digest lactose (i.e., sugars naturally occurring in milk). You're missing an enzyme—lactase—that is needed to digest milk sugar. There are lactose-free dairy products. Butter and hard cheeses (but not whey cheese) don't contain any lactose—it disappears during the manufacturing process. Heavy cream and crème fraîche contain small amounts of lactose, but much more fat. Try and see if you can tolerate them.

You'll have to stay away from dairy products if you're allergic to milk protein (which is mostly casein). However, milk fat usually causes no problems. Try ghee, which is clarified butter. Melt some butter in a pan and bring it to a boil, and then let it simmer for a short while. Skim off the frothy surface and spoon out the clear, melted fat, leaving the coagulated protein solids at the bottom of the pan. To ensure that you're not getting any milk protein, pass your clarified butter through a coffee filter. Save the butter in a jar in the refrigerator. Ghee keeps for a long time, and can be used in any recipe that calls for fat.

However, don't attempt to use ghee if you are extremely allergic to milk proteins—there might still be traces of protein molecules in the butterfat.

Cholesterol
Wouldn't you think that, after millions of years of evolution, our bodies know best how much cholesterol they can handle? There is no such thing as bad cholesterol. If you eat a nutritious diet—LCHF—you'll be as healthy as you can possibly be.

It has been proven that, on average, people who suffer a heart attack have lower levels of cholesterol than people who do not experience this type of cardiac event. This clearly shows that a risk factor for heart attack is, if anything, a low level of cholesterol.

Annika Dahlqvist, MD

$\mathcal{S}$oon I'll reach my three-year anniversary with LCHF nutrition. Before, I was searching for something that would alleviate my problems, after fibromyalgia started bothering me following a twelve-year-old work-related back injury. A lecture I attended, by the physician Annika Dahlqvist, MD, made me change my eating habits completely. Annika was free from pain and fatigue brought on by fibromyalgia, and if she was able to cure herself, then perhaps I could too. At home, our daily diet went from one based on quantities of sandwiches, pasta, cookies, and cakes, to the exact opposite. Now my shopping list reads: meat, fish, poultry, eggs, butter, heavy cream, cheese, vegetables, and berries. I started to feel a marked improvement in as little as a few weeks; today I seldom experience any aches or pains, and I feel totally energized.

My career was in the restaurant business, and my introduction to that life was as the manager of the cold foods buffet; later I progressed to prep cook, and later still to chef. After eighteen years in many different establishments, I injured my back, which effectively ended my career and professional working life. I have since lived with a disability, but I try to remain as active as I can with daily physical therapy and long walks. Over the last few years I've had to rethink my diet and have become a LCHF chef in my home kitchen. It's an exciting challenge to transform our traditional, long-cherished dishes into new favorites.

About two years ago I started writing online about my food. Today, my blog, *Birgitta Höglunds Mat* (*Birgitta Höglund's Food*), contains more than eight hundred of my recipes; I have never worked on such a varied or appetizing menu as I do now. Because I'm constantly taking notes about my food, I've had the opportunity to rework all my daily favorites, including bread recipes and desserts, into a more LCHF-oriented way of eating. I also write food columns for two newspapers. One is published in my home-town of Östersund, in the north of Sweden, in which I focus my articles predominantly on local food such as game, berries, and wild mushrooms. The other paper runs in Alanya, Turkey, and my writing centers more on easy dishes—sourcing local, Turkish produce—that can be prepared on the two-burner hotplate of a hotel room. The reason for this is that for the last nine years my partner, Lennart, and I spend most of winter in Alanya, because the climate and the Mediterranean food are so healthy for us.

My interest in natural, local food is nothing new. My mother is an excellent cook, and my father hunted and fished. We grew our own vegetables, and we picked berries and mushrooms in areas around our village, at the base of Fulu Mountain. This love of local, simply prepared food runs like a thread through my life. I thank my mother from the bottom of my heart for giving me life and for instilling in me this love of good food.

Birgitta Höglund

$\mathcal{O}$ver the last few years, I've chosen to eat foods that are as natural and as locally sourced as possible. This means organically grown, produced, and, to the extent that it's possible, pesticide-free.

It is my firm belief that we and the environment are both at our best when we focus on feeding ourselves with what can be found in nature—food that is raised the way our ancestors did it. Food animals also lead a better life if they're allowed to graze freely outside, as they've done for centuries. Our health would certainly improve if we ate food raised in this manner; in fact, it's the kind of food we ate here in the Nordic countries before artificial fertilizers, factory farming, and commercial food industries took over.

I never use any grains in my recipes. Cookies and bread derive their attractive shape and taste from ground almonds or psyllium husk powder instead of grain flour. Whole natural psyllium husk powder is the ground husk of the psyllium plant, *Plantago ovata*, and it is entirely free of carbohydrates and gluten. You can often find it in the aisle reserved for gluten-free products in grocery stores. If it's not available there, you can easily find it online. Almond flour, which is made up of ground almonds, can also be found at health-food stores and online; you can also make your own by grinding almonds in a nut grinder or food processor.

If a sauce or soup needs thickening, instead of using flour I add an egg yolk diluted with some heavy cream, and whisk it into the soup at the end of cooking. I let it simmer for a short while because it thickens the dish quickly. Often a sauce will be thick enough without having to add the yolk—all you have to do is let it cook a while longer over low heat. The flavor of a sauce that includes a hearty splash of heavy cream is so much tastier than one based on milk and flour.

If I want an added touch of sweetness to enhance the taste of a cookie or a dessert, I will occasionally use liquid honey. Organically harvested honey has nutritional value that sugar lacks; however, keep in mind that honey elevates blood sugar the same way ordinary sugar does. Usually, only a tiny drop is needed, but if you suffer from diabetes, are obese, or are addicted to sugar, I recommend that you substitute honey with erythritol or stevia. These two ingredients are natural sweeteners that do not elevate blood sugar the way honey and sugar do. All my recipes that contain a small amount of honey can easily be made completely sugar free—just swap the honey out for an equal amount of erythritol. Stevia is incredibly sweet, so only a tiny amount—as in, the tip of a knife—is required to add a sweet taste to, for example, baked goods. You might detect a very slight aftertaste of licorice, though, and it can also sometimes taste a bit bitter, so the best way to use stevia is to mix a tiny amount of it with some erythritol; using the two sweeteners together cancels out any off-putting flavors that might develop. In fact, premixed stevia and erythritol is a product—Truvia—that's easy to find at the grocery store, in the aisle next to sugar and artificial sweeteners.

Organic bouillon cubes can be found at health-food stores and they do not contain MSG, which can be bad for your health. These days, you'll also see them in most grocery stores, near regular bouillon cubes or organic foods.

I have used sugar as a preservative in some pickled items; this is the only time I use sugar. I buy organic sugar in two-pound bags. Most of the sugar stays in the pickling liquid, as the vegetables or fish don't absorb very much of it. These pickles also contain far less sugar than commercially sold pickles.

Make a point of eating fresh, locally grown berries when they're in season. Try them with some cottage cheese or heavy cream for dessert; or why not turn them into a light meal? Jams and non-alcoholic drink mixes, either unsweetened or only very lightly sweetened with your sweetener of choice, can be kept in the refrigerator or, better yet, in the freezer, to be enjoyed in winter.

These days, I don't eat fruit every day as I once did. When I do, I buy organically grown fruit, and use it with other food, or occasionally as part of a dessert. If you don't eat sweet fruit during the day, your blood sugar will tend to stay at an even and comfortable level. You won't feel any cravings for sweets, and your hunger will be synched up to when real food hits the table.

Having a small cookie with a cup of coffee is an integral part of our Nordic traditions, so I have developed several go-to recipes that are well adapted even to diets that call for fewer carbohydrates.

A variety of seeds, almonds and other nuts are not only extremely nutritious, they also make very tasty baked goods to rival old-fashioned, wheat-based cookies for an occasional treat.

An everyday accompaniment to a cup of coffee lightened with heavy cream could be a slice of cheese with some butter, or perhaps a bite of one my breads, which contain very few carbohydrates but a high amount of fat and protein. Since I started eating more natural fat and fewer carbohydrates, my portions have decreased in both size and quantity, as I feel full for longer between meals, and my blood sugar no longer takes me for a roller coaster ride.

My food budget has shrunk too, thanks to my LCHF diet, even though I now buy organic and higher-quality food.

The recipes in this book generally make 1 to 2 servings, but appetite often wanes with age, so some of you might be able to stretch a few of these dishes a little further.

Also, all of these recipes can be frozen in small containers. It's handy to be able to grab something to eat from the freezer, and quickly heat it up on days when you don't have the time, strength, or motivation to stand over a hot stove.

Birgitta Höglund

Hors d'Oeuvres and Open-Face Sandwiches

Crayfish

For us Swedes, late summer festivities are strongly linked with crayfish. While it's no longer easy to find crayfish spawned and grown locally, you can still get your hands on good quality imported crayfish for all your summer get-togethers.

Serve the crayfish with some of the sauces from p. 74. Cheese bread with caraway is also a good accompaniment (see recipe on p. 61).

Savory Salmon Paté *with Shrimp*

This paté is simple to prepare in advance, and keeps well in the refrigerator until time for dinner. You can serve this dish either as an appetizer or, if served in larger portions, as a small entrée for supper. I made it with pickled raw salmon—also called *gravad* salmon—but it's just as good when made with smoked salmon. *Gravad* salmon is a Swedish traditional dish that consists of salt and sugar-cured raw salmon.

Makes 2–4 servings

3 1/2 oz. (100 g) *gravad* or smoked salmon, sliced
2 tablespoons dill, chopped
1 tablespoon chives, chopped
1 teaspoon lemon juice, freshly pressed
1/4 cup (50 ml) crème fraîche
salt and pepper

Dice the salmon very finely. Stir the chopped herbs and lemon juice into the crème fraîche. Fold in the diced salmon, season with some salt and pepper, and mix well.

Mold the paté by pushing a few tablespoons of the salmon mixture into small, empty molds. Cover them with plastic wrap and store them in the refrigerator for about an hour, just to firm up the pate to help with unmolding. Or, shape the paté into small eggs with a spoon dipped in warm water.

Serve the paté with lettuce leaves, lemon, avocado, and peeled shrimp.

Smoked Salmon *with Egg Butter*

Finnish egg butter is typically served with Karelian pasties that are stuffed with rice. My variation is less rich in carbohydrates. I season the egg butter with smoked salmon and serve it with waffles (recipe on p. 65).

Makes 2 servings

2 eggs, hard-boiled
1 oz. (25 g) butter, room temperature
3 1/3 fl. oz. (100 ml) smoked salmon, finely diced

Peel and chop the eggs finely. Mix them with the softened butter and stir in the diced salmon. Dip a spoon in warm water and use the spoon to shape small egg shapes from the egg butter. Place the egg shapes on cooled almond waffles. Serve with lettuce, dill, and radishes.

Cottage Cheese Spread *with Tuna*

A can of water-packed tuna can turn into a truly delicious hors d'oeuvre. A blend of cottage cheese, mayonnaise, and fresh herbs brings out all facets of its flavor.

I only use water-packed tuna, since oil-packed tuna is often marinated in inferior-quality oil. If you'd like to save one half of the avocado, leave the stone in the half you're not using. This will make the leftover avocado keep longer, and also helps prevents it from turning brown.

Makes 2 servings

1 small can of water-packed tuna
1/3 cup + 1 1/2 tablespoons (100 ml) cottage cheese
1/4 cup (50 ml) mayonnaise
1 teaspoon lemon juice
1 tablespoon parsley, chopped
1 tablespoon thyme, chopped
1 tablespoon leek, chopped
salt and pepper

Leave the tuna to drain thoroughly in a sieve. Mix the cottage cheese with mayonnaise and lemon juice. Stir in the chopped herbs and leek, and season with salt and pepper.

Separate the tuna into small pieces, and fold them carefully into the cottage cheese mix. Spoon the mix into one half of an avocado, garnish with radish, lemon, and lettuce.

Any leftover tuna spread will keep for a few days in the refrigerator.

Smoked Reindeer Roast

with Horseradish Cream

The best meat we can eat includes different cuts of reindeer meat and wild game. Reindeer graze on natural food up in the *fjelds*, or Nordic mountains. This makes for very tasty and nutritious meat.

I have relatives who hunt elk. If you're not as fortunate as I am, you can still find reindeer, elk, and deer meat in certain well-stocked grocery stores. Smoked game can be found in the meat and deli sections.

Makes 2 servings

1 oz. (25 g) smoked reindeer roast
2 tablespoons finely grated horseradish
sweetener equivalent to 1/2 teaspoon of liquid honey (erythritol, stevia, or a blend of the two)
1/2 teaspoon lemon juice
salt flakes
1/3 cup + 1 1/2 tablespoons (100 ml) heavy cream

Dice the meat finely. Mix the horseradish with chosen sweetener, lemon, and salt—this will enhance the flavor of the radish.

Whip the cream to soft peaks. Stir the horseradish mix into the cream. Fold in the diced meat.

Serve on small plates and garnish with lettuce leaves, finely julienned leek, lingonberries (or cranberries), and a few thin slices of apple.

Open-Face Sandwich

with Hamburger and Sautéed Onion

Hamburger with onions is an old-time, homey favorite. Here I've remade it into an open sandwich using my homemade seed bread. For those of you who eat LCHF, this is so much better than the classic side of skillet potatoes (or french fries).

Makes 1 serving

1 yellow onion
2 tablespoons butter
salt and pepper
2 small cooked hamburgers (see recipe on p. 130)
1 slice of seed bread (see recipe on p. 58)

Slice the onion finely and sauté in butter until soft, without browning it too much; season lightly with salt and freshly ground pepper. Heat the hamburgers with the onions. Serve the sandwich on seed bread with lingonberries (you can substitute with cranberries if you can't find lingonberries) and parsley.

Open-Face Sandwich *with Liver Paté,*

Bacon, and Onion

Two classic fillings for Danish open-face sandwiches are roast beef and liver paté. Made this way, these sandwiches make a small yet complete meal.

Use one of my homemade breads, which contain few carbohydrates, to turn them into nutritious sandwiches as well. Bread recipes can be found on p. 58.

The second sandwich in the picture contains a layer of roast beef and Danish rémoulade sauce (see recipe on p. 74).

Makes 1 serving

2 slices of bacon
1 small onion
1 tablespoon butter + more for spreading
2 thick slices of liver paté
1 slice of almond bread

Slice the bacon and onion finely; sauté in a tablespoon of butter until the bacon is crispy and the onion slightly browned.

Set the liver paté on the buttered bread, and heap the onion and bacon on top. Garnish with lettuce leaves and parsley. Thinly sliced, pickled cucumbers make a nice accompaniment (see recipe on p. 78).

Meatballs *with Beet Salad on Seed Bread*

One of the most popular open-face sandwiches in Sweden's cafés is round wheat bread with meatballs and beet salad. It's very tasty, but it's not on the menu if you eat according to LCHF.

You can put together a far better sandwich by using my seed bread as a starting point—you'll end up with a more nutritious sandwich that will keep you full much longer. The recipes for beet salad and homemade meatballs are on p. 78 and p. 130, respectively. My seed bread recipe is on p. 58.

Makes 1 serving

butter
1 slice of seed bread
1/4 cup (50 ml) beet salad
3–4 meatballs, sliced in half

Butter the bread. Place a dollop of beet salad and the meatballs on the buttered bread. Garnish with lettuce, parsley, and thin slivers of apple.

Smoked Turkey Roulades

with Pear-Studded Cottage Cheese Filling

Cottage cheese, infused with the inviting taste of fresh pears and herbs, is the filling for these slices of smoked turkey. This simple yet easy, light meal can be made more substantial by adding a boiled egg. The roulades can also be made with smoked or cooked ham.

Makes 1–2 servings

1 small pear
1/4 cup (50 ml) cottage cheese
1 tablespoon crème fraîche
1 tablespoon mayonnaise
1 tablespoon lemon balm, finely chopped
1 tablespoon parsley, finely chopped
lemon juice
salt and pepper
4 slices of smoked turkey

Peel and dice the pear into small cubes. Stir together cottage cheese, crème fraîche, and mayonnaise in a bowl.

Mix in the finely chopped herbs; season with lemon juice, salt, and freshly ground pepper. Carefully fold in the diced pear.

Place a dollop of cottage cheese on each slice of meat and roll it up.

Norwegian Morning Coffee

A regular snack that I enjoy with my morning coffee is a slice of homemade, low-carb bread topped with authentic Norwegian goat cheese and a slice of full-fat cheese.

I was born and raised only a few miles from the Swedish–Norwegian border; consequently, the brown, Norwegian goat's whey cheese has been a big favorite of mine since childhood. The whey cheese does contain quite a lot of milk sugar, but it also contains valuable nutrients from the goat milk: the protein content is 12 percent, the fat content is 27 percent, and the carbohydrate content is 38 percent. It also has a fairly high amount of vitamins and minerals.

The whey cheese, when paired with a slice of full-fat cheese on buttered seed bread, doesn't rack up high levels of carbohydrates, and the taste is simply divine when accompanied by a cup of coffee.

Jellied Veal Brawn Aspic with Beet Salad and Hard-Boiled Egg

Veal brawn in aspic (also called headcheese) used to be a butcher shop or delicatessen standard. Today it's experiencing a renaissance, and many small, local food artisans and boutiques are keeping the tradition alive.

This picture shows a very good reindeer calf brawn, produced by a new generation of food artisans. I feel that it's critical to promote and eat locally sourced food instead of only buying mass-produced goods from giant food corporations. Indeed, prices for these items might be slightly higher, but the superior quality and flavor are worth every extra penny.

Makes 1 serving

3 slices of veal brawn in aspic
1/4 cup (50 ml) beet salad (see recipe on p. 78)
1/2 hard-boiled egg

Set the veal brawn and the beet salad on a plate along with some slices of egg. A sprig of parsley enhances the flavor and adds a nice touch of color.

Gentleman's Delight

Among Nordic fish dishes that feature herring, Gentleman's Delight is a classic. It's lovely by itself with some lettuce, or on a slice of seed bread (see recipe on p. 58).

4 pickled herrings (sprats) or anchovy filets
1 egg, hard-boiled
1 tablespoon crème fraîche
1 tablespoon smoked caviar spread (or canned fish roe)
1 tablespoon chives, chopped
1 tablespoon dill, chopped
1 tablespoon finely chopped yellow onion

Place the sprats or anchovies in a colander to drain. Dice the egg into small cubes. Mix the crème fraîche with the roe, herbs, and onion.

Dice the sprats or anchovies and fold them into the crème mix, together with the diced egg.

Bread, Pancakes, and Breakfast

LCHF Bread and Butter House

Ordinary bread has no place in LCHF nutrition. However, if you'd like to enjoy a tasty sandwich from time to time, it's easy to bake bread from other ingredients besides grain flour.

I have made three different breads with variations in seasoning. They're great as-is with only some butter and cheese, or as part of a small meal with more substantial fixings.

The butter house used to decorate the dinner table at Christmas and on other holidays. It wasn't enough that the butter taste good, it had to be beautiful to look at, too. A special, richly ornate wooden mold is soaked in water overnight, then assembled the next day. Softened butter is inserted into the mold and left to firm up until the next day. The mold is then opened up to reveal a true piece of art—made of butter.

SEED BREAD

Makes 12 buns

1 oz. (25 g) organic coconut fat
1/3 cup + 1 1/2 tablespoons (100 ml) almonds, ground
2 tablespoons sesame seed
2 tablespoons flax seeds
1 tablespoon psyllium husk powder
2 teaspoons fennel seeds, coarsely ground
2 teaspoons anise seeds, coarsely ground
1 teaspoon baking powder
1 teaspoon salt
2 large eggs
1/3 cup + 1 1/2 tablespoons (100 ml) crème fraîche

Preheat the oven to 400°F (200°C). Line a baking sheet with parchment paper.

Melt and then cool the coconut fat.

Mix all the dry ingredients in a bowl. Add the eggs, crème fraîche, and the cooled coconut fat. Mix to a smooth batter, and allow to rise for 5 minutes.

With a spoon, place mounds of the batter on the prepared baking sheet. Flatten the mounds to form the shape of round cakes.

Bake for about 12–15 minutes, until the bread has browned slightly. Let cool on a baking rack and then store in the freezer. Defrost the bread in the toaster for that freshly baked taste.

CHEESE BREAD
Makes 12 buns

1 oz. (25 g) butter
2 large eggs
1/3 cup + 1 1/2 tablespoons (100 ml) crème fraîche
3/4 cup + 2 tablespoons (200 ml) Parmesan* cheese (or Västerbotten), grated
1/4 cup (50 ml) pumpkin seeds
1/4 cup (50 ml) sunflower seeds
2 tablespoons psyllium husk powder
2 teaspoons caraway seeds
1/2 teaspoon salt
sesame seeds, for sprinkling

Preheat the oven to 400°F (200°C). Line a baking sheet with parchment paper.

Melt the butter and let it cool. Using two bowls, separate the eggs into yolks and whites. Mix the yolks with the crème fraîche and the melted butter.

Stir in grated cheese, seeds, psyllium husk powder, caraway seeds, and salt. Beat the egg whites until stiff peaks form; fold them into the batter.

Using two spoons, place rounds of batter on the prepared baking sheet. Sprinkle some sesame seeds on top of the rounds, and bake them for about 12–15 minutes.

Let cool on wire rack. Freeze in bags. Defrost as needed in a barely warm oven, or heat for a short time in the microwave.

*Go easy on the salt if using grated Parmesan cheese.

ALMOND BREAD
Makes 10 buns

1 3/4 oz. (50 g) butter
3/4 cup + 2 tablespoons (200 ml) almond meal
2 large eggs
1/4 cup (50 ml) heavy cream
1/2 teaspoon salt
2 tablespoons psyllium husk powder
1 teaspoon baking powder

Preheat the oven to 435°F (225°C) and line a baking sheet with parchment paper.

Melt the butter in a saucepan and let it cool a little. Stir in almond meal, eggs, heavy cream, and salt.

Mix the psyllium husk powder with baking powder and add to the batter. Mix thoroughly and let sit for about 5 minutes.

Place dollops of the batter on the prepared baking sheet and spread it into oval buns.

Bake for about 10–12 minutes. Let cool on wire rack. Freeze the bread and defrost as needed in a toaster or the oven.

Galettes *(Swedish Pancakes)*

Galettes, or Swedish pancakes, are delicious, easy to make, and really nutritious to boot. This recipe has been adapted to fit an LCHF nutritional profile featuring few carbohydrates.

Instead of flour, the galettes contain psyllium husk powder, which have no carbohydrates. Look for them in the gluten-free section of your grocery store or health-food store. Several merchants also sell psyllium husk powder online.

Psyllium husks swell to many times their original size when they come in contact with liquids, and hold the batter together.

Makes 5–6 small pancakes

2 large eggs
1/3 cup + 1 1/2 tablespoons (100 ml) heavy cream
1/8 teaspoon salt
1 tablespoon psyllium husk powder
butter, for frying

Beat the eggs with the cream and salt. Sprinkle the psyllium husk powder over the egg batter while continuing to beat. Leave the batter to rise and swell for 10 minutes.

Whisk the batter thoroughly and fry the galettes in a non-stick, Teflon-style frying pan (to make it easier to flip them).

Brown a dollop of butter and use about 1/4 cup (50 ml) of batter per pancake. Fry them for a few minutes on each side until they're golden brown and crisp around the edges.

Whisk the batter thoroughly between cooking each galette. If the batter thickens too much, dilute it with some added water or cream.

Almond Waffles

Waffles make a delicious light breakfast or evening meal. A waffle is also an eminently suitable substitute for bread as part of an appetizer, or with a topping to accompany a cup of coffee.

When I went berry picking as a kid, we often had buttered waffles with cheese and goat cheese for a picnic. Since waffles are not only very tasty but also portable, they're excellent to bring along on trips.

Makes 1 large waffle or 2 small waffles

1 large egg
1/4 cup (50 ml) heavy cream
1 teaspoon psyllium husk powder
1/4 cup (50 ml) almond meal
2 teaspoons butter, melted + more for brushing
1/2 teaspoon baking powder
1/8 teaspoon salt

Beat the eggs and cream together; sprinkle in the psyllium husk powder and almond meal while continuing to beat. Mix in the melted butter, and season with salt. Brush the waffle iron liberally with melted butter.

Keep the waffle iron closed before adding the batter to let it heat up properly. The batter will be quite thick, so use a spoon to spread it over the hot iron.

The waffle will be ready when it's golden brown. Serve with whipped cream and cloudberries, or any berries of your choice.

Oven Pancake

Studded with Hot-Smoked Pork Belly Croutons

This oven pancake is a perfect example of hearty home fare; substituting almond meal and psyllium husk powder for wheat flour has made my version of this classic even more satisfying and nutritious. Many hours will pass before you're ready for another meal after eating this stick-to-your-ribs dish.

Makes 4–6 servings

4 large eggs
3/4 cup + 2 tablespoons (200 ml) whole milk
1/3 cup + 1 1/2 tablespoons (100 ml) heavy cream
1/2 teaspoon salt
3/4 cup + 2 tablespoons (200 ml) almond meal
1 1/2 tablespoons psyllium husk powder
5 1/4 oz. (150 g) hot-smoked pork belly
butter for the pan

Preheat the oven to 435°F (225°C). In a bowl, beat together the eggs, milk, cream, and salt. Stir in almond meal and psyllium husk powder, and mix thoroughly. Leave the batter to rise for about 10 minutes.

Dice the pork belly and scatter into a small, buttered baking pan. Set the pan in the oven and cook it until the pork has got some color. Carefully pour the pancake batter over the pork. Bake on the middle rack of the oven for about 20–25 minutes, until the pancake is golden brown. Serve it with fresh uncooked jam made from lingonberries and sweetener (cranberries work well, too).

Freeze the leftovers, and reheat them in the oven on a day when cooking is not at the top of your to-do list.

Muesli

Your breakfast needn't include bacon and eggs or omelets, day in and day out, simply because you follow an LCHF diet.

A bowl of yogurt with some homemade muesli is a great alternative for those who don't really care for a cooked breakfast. Recently, several different kinds of high-fat yogurts have been introduced and are sold in well-stocked and specialty grocery stores. Look for Turkish and Greek yogurt with at least 10 percent fat; there's also Russian yogurt with 17 percent fat. They are all very mild and taste wonderful.

To make it a bit more special, try adding a spoonful of cottage cheese and heavy cream to the yogurt. This will increase the protein content, too, which will help you stay satiated for a long time.

2 tablespoons butter
2 tablespoons organic coconut fat
3/4 cup + 2 tablespoons (200 ml) flax seed
1/3 cup + 1 1/2 tablespoons (100 ml) sunflower seed
1/3 cup + 1 1/2 tablespoons (100 ml) sesame seed
1/3 cup + 1 1/2 tablespoons (100 ml) unsweetened shredded coconut
1 tablespoon honey, or sweetener of your choice
2 tablespoons cinnamon
1/8 teaspoon salt

Melt the butter and coconut fat in a skillet, and let it brown slightly.

Pour in the seeds and the shredded coconut, and stir well. Sauté the seeds and coconut over low heat for a few minutes, stirring constantly to avoid burning it.

Drizzle in the honey—or sweetener of your choice—and season with cinnamon and salt. Let the mix cook a little longer, while stirring continuously.

Let the muesli cool, and then pour it into a jar.

Serve with yogurt and berries.

Buckwheat Porridge

Buckwheat, contrary to what its name seems to imply, is entirely gluten-free. It is not a grass like ordinary grains, but pseudocereal. It has long been a food staple in Eastern Europe, and it has good nutritional value with a high percentage of protein and vitamin B. Note that buckwheat is high in carbohydrates, so the porridge is not a good choice if you are a diabetic. The buckwheat can be replaced with 2 tablespoons almond meal and 1 tablespoon sesame seeds.

An evening bowl of porridge, topped with a dollop of melted butter, a pinch of cinnamon, a splash of whole milk, and perhaps a few blueberries, will settle you in for a good night's sleep.

Makes 1–2 servings

1/4 cup (50 ml) crushed buckwheat groats (available in health-food stores)
2/3 cup (150 ml) whole milk
1/8 teaspoon salt
1/4 cup (50 ml) heavy cream
1 tablespoon butter

Mix the buckwheat groats, 1/4 cup (50 ml) of the whole milk, and salt in a thick-bottomed saucepan. Bring to a boil while stirring constantly. Mix in the rest of the milk, the cream, and the butter.

Bring the mix back to a boil, and let it simmer for 15 minutes over low heat. Stir now and then to prevent the porridge from burning.

Cold Sauces and Tasty Sides

SAUCES – P. 74

Danish Rémoulade Sauce
Dill Mayonnaise
Nobis Sauce
Mustard Sauce

FLAVORED BUTTERS – P. 77

Parmesan Butter
Caviar Butter with Dill
Lemon Butter
Mustard Butter

PICKLES – P. 78

Pickled Beets
Pickled Cucumbers
Beet Salad

Uncooked Lingonberry Preserves – p. 81

Black Currant Jelly – p. 82

Sauces

A creamy sauce enhances many dishes while also making them more nutritious and satisfying. These sauces, using mayonnaise and crème fraîche as a base, are excellent complements to both meat and fish. In the summer, several of them are good to serve with crayfish or a dinner salad. For best results, prepare them an hour ahead and let them rest in the refrigerator before serving to let the flavors develop fully. The sauces will keep up for up to a week if refrigerated.

DANISH RÉMOULADE SAUCE

1/3 cup + 1 1/2 tablespoons (100 ml) finely diced pickled cucumber

1/3 cup + 1 1/2 tablespoons (100 ml) finely diced apple

1 tablespoon capers

1/3 cup + 1 1/2 tablespoons (100 ml) mayonnaise

1/4 cup (50 ml) heavy cream

1/4 cup (50 ml) crème fraîche

salt

1 teaspoon curry powder

Dice the cucumber and apple. Finely chop the capers. Whisk together the mayonnaise, heavy cream, and crème fraîche. Fold the vegetables and the apple into the mayonnaise and season with salt and curry powder.

DILL MAYONNAISE

1/3 cup + 1 1/2 tablespoons (100 ml) mayonnaise

1/3 cup + 1 1/2 tablespoons (100 ml) crème fraîche

2 tablespoons dill, finely chopped

1 teaspoon lemon juice

1/2 teaspoon Dijon mustard

salt and pepper

Mix mayonnaise and crème fraîche. Stir in dill, lemon juice, and mustard. Season with salt and freshly ground pepper.

NOBIS SAUCE

1 soft-boiled egg

1 teaspoon unsweetened mustard

1 tablespoon vinegar

3/4 cup + 2 tablespoons (200 ml) mild-tasting olive oil

2 tablespoons chives, finely chopped

1 small clove garlic, pressed

1/2–1 teaspoon salt

1/4 teaspoon pepper

Peel the soft-boiled egg and mash it thoroughly in a bowl with a fork. Mix in the mustard and vinegar. Whisk in the oil drop by drop at the start, and then add in a thin stream. Whisk vigorously until the sauce has thickened.

Fold in chives, garlic, and salt and pepper.

MUSTARD SAUCE

1/3 cup + 1 1/2 tablespoons (100 ml) mayonnaise
2 teaspoons unsweetened mustard
1 teaspoon honey or sweetener equivalent
1/4 teaspoon white vinegar (or 12 percent acetic acid)
salt and pepper
1/3 cup + 1 1/2 tablespoons (100 ml) whipped cream

Mix mayonnaise, mustard, and honey or sweetener of choice. Season with the vinegar, salt, and freshly ground pepper. Carefully fold in the whipped cream.

Flavored Butters

Flavored butters are very easy to make, and their flavors complement many dishes.

With a broad-bladed knife, shape the butter into a log in a piece of wax paper. Dot the butter along the midline and fold one side of the paper over the butter. Draw the knife along the butter to pack it into a nice log shape. Twist each end of the paper so it looks like a Christmas cracker. Store the log in the freezer or refrigerator. You can also place the butter in a piping bag; pipe small roses on a baking sheet and freeze them individually before placing them in a freezer bag or box. Place the roses at room temperature a good amount of time before serving to let the flavors develop fully.

PARMESAN BUTTER

3 1/2 oz. (100 g) butter, room temperature
1/3 cup + 1 1/2 (100 ml) tablespoons grated Parmesan (or Västerbotten) cheese
salt and pepper

Whip butter and cheese in a bowl—the easiest way to do this is with a handheld electric mixer. Season with salt and freshly ground pepper.

CAVIAR BUTTER WITH DILL

1 yolk from a soft-boiled egg
3 1/2 oz. (100 g) butter, room temperature
2 tablespoons Swedish smoked caviar spread (available at IKEA, or online Swedish specialty stores), or a small can of fish roe
2 tablespoons dill, chopped

Peel the egg and remove the soft-boiled yolk. Mash the yolk with a fork and mix it together with the butter and the caviar spread. Stir in the dill.

LEMON BUTTER

juice from 1/2 lemon
3 1/2 oz. (100 g) butter, room temperature
salt and pepper

Squeeze the lemon juice over the butter and season carefully with salt and freshly ground pepper. Whip the butter with a handheld electric mixer until light and airy.

MUSTARD BUTTER

3 1/2 oz. (100 g) butter, room temperature
1 tablespoon parsley, chopped
1 tablespoon chives, chopped
2 teaspoons unsweetened mustard
salt and pepper

Mix all ingredients in a bowl and whip vigorously with a handheld electric mixer. Season with salt and freshly ground pepper, and adjust for more herbs if desired.

Pickles

We've been canning and pickling food for a long time now. With acetic acid or vinegar, sugar, and salt, vegetables could be put up and overwintered, stocked in basements and pantries. Old-fashioned pickled beets didn't have near the amount of sugar included in today's commercial varieties.

PICKLED BEETS

2.2 lbs. (1 kg) raw beets, even in size and shape
3/4 cup + 2 tablespoons (200 ml) vinegar (or 12 percent acetic acid)
1 1/4 cups (300 ml) water
sweetener equivalent to 2/3 cup (150 ml) granulated sugar
10 whole cloves
5 white peppercorns
1 small piece of horseradish

Don't peel the beets; leave the leaves on and the root at the bottom. Clean the beets thoroughly with a bristle brush, and cook them to semi-soft in lightly salted boiling water. Rinse the boiled beets in cold water. Cut off the leaves, stalk, and root, and rub the beets' skin to remove it. Mix all the ingredients (except the horseradish) to make the pickling liquid, and bring to a boil. Add the beets to the liquid, and boil for 15 minutes. Put the beets into thoroughly cleaned glass jars, and pour in the pickling liquid. Make sure the pickling liquid covers all the beets in the jars. Cut the horseradish into small dice and add to the jars. Screw the lid on immediately, and store in the refrigerator.

PICKLED CUCUMBER

2 tablespoons vinegar (or 12 percent acetic acid)
1/3 cup + 1 1/2 tablespoons (100 ml) water
sweetener equivalent to 1–2 tablespoons honey
1/8 teaspoon freshly ground white pepper
1 hothouse cucumber
1 teaspoon salt
2 tablespoons chopped parsley or dill

Make a pickling liquid by mixing the vinegar or acetic acid, water, sweetener, and white pepper until the sweetener has completely dissolved. Slice the cucumber thinly, and layer the slices with salt in a deep plate. Cover with another plate to press onto the cucumber slices. Place a heavy weight on the top plate—a marble mortar and pestle, for example. Let the slices stand for about half an hour, and then pour off the accumulated liquid. In a glass jar, mix the cucumber slices with the pickling liquid and chopped herbs. Let the pickled cucumbers rest for a few hours in the refrigerator to allow the flavors to develop fully.

BEET SALAD

3/4 cup + 2 tablespoons (200 ml) diced pickled beets
1/4 cup (50 ml) diced pickled cucumber
1/4 cup (50 ml) diced apple
1/3 cup + 1 1/2 tablespoons (100 ml) mayonnaise
1/4 cup (50 ml) crème fraîche
1 teaspoons unsweetened mustard
salt and pepper

Place a sheet of wax paper paper on the cutting board to protect it from the red juice from the beet. Finely dice pickled beets and cucumber. Leave the vegetables to drain thoroughly in a colander. Peel and dice the apple. Mix mayonnaise, crème fraîche, mustard, and salt and pepper. Carefully fold the diced vegetables and apple into the mayonnaise sauce. Place the salad in the refrigerator for a little while before serving—it will thicken as it rests in the cold. If you want it to be a bit more liquid, add some more mayonnaise.

Uncooked Lingonberry Preserves

Lingonberries and other types of jellies are traditional sides to many Nordic homey dishes. I've decreased the traditional amount of sweetener considerably to make them fit a LCHF lifestyle.

Lingonberries keep extremely well, and therefore only need a small amount of sweetening.

2 cups (500 ml) lingonberries (or cranberries), fresh or frozen
sweetener equivalent to 2 tablespoons honey

Mix uncooked lingonberries and sweetener in a bowl. Mash the lingonberries lightly with a wooden spoon to bring out some of the juice; leave at room temperature until the next day. Stir the mix occasionally during this time. Pour the contents of the bowl into thoroughly cleaned glass jars and store in the refrigerator.

Black Currant Jelly

Jelly typically contains lots of sugar, so I have developed a new recipe here that uses gelatin to achieve the right consistency.

This jelly does not keep as well as a sugar-filled jelly, so a good way to store it is to freeze it in small jars and defrost as needed.

3 sheets of gelatin (2 1/4 teaspoons in powder)
2 cups (500 ml) fresh black currants
1/4 cup (50 ml) water
sweetener equivalent to 1–2 tablespoons honey

If using gelatin sheets, soak the sheets in cold water for 10 minutes. If using powder, follow instructions on the package.

Bring black currants and water to a boil, and simmer for a few minutes. Using a ladle, press the berries through a sieve. Mix the gelatin into the warm liquid; season with the sweetener.

Pour the jelly into small jars. Store in refrigerator, or freeze.

If the jelly is kept frozen, it might need a quick reheating and re-cooling to reach the right consistency.

Soups

Broccoli Soup *with Shrimp and Cheese*

Shrimp and other shellfish are nutritious and very rich in minerals. The saltiness from the shrimp is a nice counterpoint to the mild and creamy broccoli soup. For that extra touch, I've added some Parmesan cheese.

The soup is especially nice if you pass it through a blender or food processor, but you can also simply mash the broccoli with a potato masher right there in the saucepan.

Makes 2 servings

8 3/4 oz. (250 g) broccoli
1 small yellow onion
3 tablespoons butter
salt and pepper
1/4 organic stock cube
3/4 cup + 2 tablespoons (200 ml) whole milk
1/3 cup + 1 1/2 tablespoons (100 ml) heavy cream
1/4 cup (50 ml) grated Parmesan (or Västerbotten) cheese
3 1/2 oz. (100 g) peeled, cooked shrimp

Cut the broccoli into smaller florets and pieces, and chop the onion finely. Lightly brown the butter in a soup pot, add the vegetables, and sauté them to sweat, not to brown.

Season lightly with salt and freshly ground pepper, and add the stock cube. Add milk and cream, and let the soup come to a boil while stirring continuously.

Lower the heat and let the soup simmer for 30 minutes. Mash the vegetables, or blend them with an immersion blender; stir in the cheese and shrimp. Let the shrimp get warm—not hot, or they will get tough.

Serve each portion with half a hard-boiled egg.

Fish and Shellfish Soup

Fish soup is one of my favorite meals. Maybe it's because my mother is from a fishing family in the Finnish archipelago of Åboland, and she often prepared fish soup when I was a kid. Back then fish came from rivers and lakes in Sweden, at the foot of the northern Dalecarlia Mountains.

My soup gets its mouth-watering aroma from anise and fennel. It's also extra creamy from the inclusion of heavy cream.

Makes 2–3 servings

1 organic vegetable stock cube
1 2/3 cups (400 ml) water
1/3 cup + 1 1/2 tablespoons (100 ml) carrot, julienned finely
1/4 cup (50 ml) parsnip, julienned finely
1 yellow onion, chopped finely
1 clove garlic, pressed
1 tablespoon tomato purée
1 teaspoon anise, crushed
1 teaspoon fennel, crushed
1 teaspoon dried French tarragon
2 bay leaves
3/4 cup + 2 tablespoons (200 ml) heavy cream
salt and pepper
4 1/2 oz. (125 g) of fresh salmon
3 1/3 fl. oz. (100 ml) shrimp, peeled
3 1/3 fl. oz. (100 ml) crawfish tails

Dilute the stock cube with water in a soup pot and boil the carrot, parsnip, onion, and garlic for 10 minutes. Stir in tomato purée, anise, fennel, French tarragon, and bay leaves. Pour in the cream and bring the soup to a boil. Season with salt and pepper. Turn down the heat and let it simmer for 30 minutes.

Meanwhile, dice the salmon finely. Toward the end of the 30-minute simmer, carefully add the salmon and shellfish to the soup. Let it all cook just enough to heat the fish and shellfish through.

Soupy Stew with Salt Pork and White Cabbage

A soupy stew made with salt pork and slow-braised white cabbage and carrots is a real old-time classic. Soups cooked this way have had pride of place in Nordic cuisine for many centuries.

The pork can be swapped for elk, chuck roast, or bone-in lamb. Bone marrow adds even more flavor to the soup, and boosts its nutritional value into the bargain.

Makes 3–4 servings

1/2 small white cabbage
1 large carrot
4-inch (10-cm) leek
8 3/4 oz. (250 g) salt pork
4 1/4 cups (1 liter) water
2 bay leaves
10 white peppercorns
1 teaspoon caraway seeds
1 tablespoon apple cider vinegar
1 organic meat stock cube

Chop the cabbage into uneven pieces and slice the carrot. Slice the leek coarsely.

Place vegetables and pork in a Dutch oven or stew pot. Pour in water to cover the contents of the pot, and bring it to a boil. Skim off the foam as it rises.

Add the spices and vinegar; crumble in the stock cube. Cover the pot and let the soup simmer on low heat for 1 ½–2 hours.

Test the meat with a fork for doneness. When it's ready, remove the meat from the pot and cut it into smaller chunks. Place a few pieces of pork and some vegetables in a deep soup bowl and ladle on some of the clear liquid. A dollop of unsweetened mustard mixed with some crème fraîche adds a tasty tang to the stew.

Spinach Soup *with Hot-Smoked Pork Belly*

Spinach soup with half a hard-boiled egg is a perennial favorite of many. Here it's rounded out with some hot-smoked pork belly to make it more filling.

Spinach is loaded with nutritious vitamins and minerals; what's more, it's a delicious vegetable that we should make a point of eating far more often. I use whole spinach leaves here, which give the soup more flavor than the chopped variety.

1 small yellow onion
8 3/4 oz. (250 g) frozen whole spinach
1 3/4 oz. (50 g) hot-smoked pork belly
2 tablespoons butter
1/2 organic stock cube
pepper
grated nutmeg
3/4 cup + 2 tablespoons (200 ml) whole milk
2/3 cup (150 ml) heavy cream

Chop the onion finely and cut the spinach into smaller pieces. Cut the pork belly into thin strips, and, in a Dutch oven or stew pot, brown the strips in butter until they're just starting to develop some color. Add the onion and spinach, and sauté briefly.

Season the onion, spinach, and pork belly with the crumbled stock cube, pepper, and nutmeg. Pour in the milk and cream, and bring to a boil. Let the soup simmer for about 30 minutes, stirring occasionally. Taste and adjust for seasoning.

Top each serving of soup with half a boiled egg.

Fish on the Menu

Boiled Brook Trout

My mother has always enjoyed cooking, and she has given me many tips for recreating my old favorites.

She often prepares brook trout and other mountain river fish, and they turn out amazingly moist and delicious. Below you'll discover a very traditional and simple way of preparing fish. If you can't get your hands on brook trout, feel free to substitute another fish.

Makes 1 serving

1 carrot
4-inch (10-cm) piece of leek
1 brook trout
water, to barely cover the fish
2 bay leaves
5 allspice berries
5 white peppercorns
a few sprigs of parsley
2–3 teaspoons salt

Thinly slice the carrot and the leek. Rinse the fish thoroughly and place it in a pot. Add water, vegetables, and spices. Bring to a boil, reduce heat, and simmer covered for about 20 minutes.

With a slotted spoon, remove the fish from the pot and serve it with the cooked vegetables and some dill-flavored mayonnaise (see recipe on p. 74).

Boil the fish's cooking water to reduce it to half the original amount. Freeze the resulting stock in small jars or in an ice-cube tray. These make good flavor enhancers in fish soups and sauces.

Oven-Baked Smoked Baltic Herring

This little kipper box was often served at smorgasbord or at dinner parties in the 1950s to 1960s. It is both easy to prepare and tasty.

Makes 1–2 servings

2 tablespoons butter + extra for buttering dish
1 small smoked Baltic herring
3 fresh mushrooms
salt and pepper
2/3 cup (150 ml) heavy cream
2 tablespoons dill, finely chopped

Preheat the oven to 400°F (200°C). Butter a small ovenproof dish.

Clean and bone the fish; break it into smaller pieces and place them in the ovenproof dish. Slice the mushrooms and fry them in butter until golden brown; season lightly with salt and freshly ground pepper. Cover the fish with the fried mushrooms. Whip the cream with the chopped dill and pour over the fish and mushrooms.

Bake for 15 minutes, or until the top of the fish has turned into a nicely colored gratin. Serve with grated carrot or a green salad.

Sweet and Salt-Cured Salmon

with Scrambled Eggs and Chives

Occasionally, you just want to eat something quick and simple. That doesn't mean you have to skimp on nutrition and flavor, however, which makes scrambled eggs with a side of cured salmon a great choice for a healthy light meal. The fat and protein from the fish and eggs will keep you satiated and happy for a long time.

Makes 1 serving

2 large eggs
2 tablespoons butter
1/4 cup (50 ml) heavy cream
1 tablespoon chives, chopped
salt and pepper
2–3 slices cured salmon

Whisk the eggs thoroughly. Brown the butter in a pan, and stir in the eggs. Add the cream and heat carefully while stirring. Let the mixture simmer over low heat for a few minutes while stirring constantly to make the eggs creamy. Season with chopped chives, salt, and freshly ground pepper. Serve with a few slices of cured salmon.

Hot-Smoked Baltic Herring

with Hard-Boiled Eggs and Mayonnaise

You won't easily find anything simpler or more appetizing than smoked Baltic herring with hard-boiled eggs. This meal's high level of omega-3 makes it a real nutritional gem, too. Choose organic eggs, as they taste better and they're of better quality; the hens have been left to freely roam and feed outside, which is their natural way of eating.

Makes 1 serving

1 hard-boiled egg
2 tablespoons mayonnaise
1 small smoked Baltic herring

Peel and cut the egg in half. Pipe in or add on a dollop of mayonnaise, and serve with the smoked herring. Tomato, cucumber, and dill are all good sides.

Oven-Cooked Cod with Egg Sauce

Boiled cod with a creamy egg sauce is not merely delicious, it's also very easy to prepare. The fish is cooked in the oven to ensure that all taste and nutrition are preserved. A side of asparagus wrapped in bacon enhances both the flavor and color of the dish.

Makes 1–2 servings

5 1/4 oz. (150 g) cod fillet
salt and pepper
1 tablespoon butter, melted + extra for frying
1 slice of bacon
3–4 stalks asparagus, cooked

Preheat the oven to 350°F (175°C). Butter an oven-proof dish.

Season the cod fillet all over with salt and freshly ground pepper. Set the fish in the buttered dish and brush with melted butter. Cover the dish with heavy-duty foil.

Bake for 10–15 minutes, depending on the thickness of fish. You should serve it warmed through—not more—or it will dry out very quickly.

Wrap the slice of bacon around a bunch of cooked asparagus. Brown it in a dollop of butter and serve with the cod and egg sauce, recipe following.

EGG AND PARSLEY SAUCE

1/3 cup + 1 1/2 tablespoons (100 ml) heavy cream
1/4 organic vegetable stock cube
salt and pepper
1 hard-boiled egg, diced
1 tablespoon parsley, chopped

Peel and dice the egg. Bring the cream to a boil in a saucepan and crumble in the bouillon cube. Let it simmer a little until the sauce thickens; taste and season with salt and freshly ground pepper. Carefully fold in diced egg and chopped parsley.

Oven-Baked Baltic Herring

with Onion Sauce

Our traditional food is deeply rooted in me. One of my favorites is fried herring with onion sauce. This is an easy way to cook herring with a minimum of work. Instead of frying it on the stove, it can simmer directly in onion sauce in the oven.

Makes 2 servings

1 yellow onion
3 tablespoons butter
2 salted herring fillets, soaked 8 hours or overnight in water
pepper
2/3 cup (150 ml) heavy cream

Preheat the oven to 435°F (225°C). Butter an oven-proof dish.

Slice the onion and brown it lightly in two tablespoons of the butter. Let it cook gently until soft. Place the herring in the buttered dish and place a few pats of butter on top. Place the fried onion all around the fish; season with some freshly ground pepper. Pour the cream over the onion.

Bake for about 15 minutes, until the fish has a nice color and the cream has thickened.

Serve the herring with cooked beets, carrots, cauliflower, or broccoli.

Cauliflower *in LCHF Dill Béchamel*

To make an LCHF version of potatoes cooked in dill—a dish that's traditionally served with salmon—I have replaced the potatoes with cauliflower, which has far fewer carbohydrates and is also very tasty when cooked in béchamel. You can vary the flavor by adding mustard, horseradish, parsley, or other herbs.

Makes 2–3 servings

2 cups (500 ml) cauliflower, cut into florets
3/4 cup + 2 tablespoons (200 ml) heavy cream
1 organic vegetable stock cube
2 tablespoons dill, chopped finely
salt and pepper

Cook the cauliflower in lightly salted water until al dente. Drain well. Put the cauliflower back in the saucepan and add the cream.

Bring to a boil and add the bouillon cube and spices. Let simmer until the cauliflower is soft and the cream has thickened. Stir in the chopped dill, and serve the cauliflower béchamel with sautéed salmon.

Salmon Gratin *with Cauliflower*

The potato, our most starchy root vegetable, wreaks havoc on blood sugar, which spikes after eating a meal with potatoes. To keep blood sugar in check during the day, make it a habit to opt for vegetables that grow above ground.

The taste of cauliflower is mild and neutral, so it works very well as a substitute for potatoes in a salmon gratin.

Makes 2 servings

1 small head of cauliflower
1/3 cup + 1 1/2 tablespoons (100 ml) whole milk
2 large eggs
1/3 cup + 1 1/2 tablespoons (100 ml) heavy cream
salt and pepper
3 1/2 oz. (100 g) cured salmon
2 tablespoons dill, finely chopped

Preheat the oven to 350°F (175°C). Butter an oven-proof dish.

Separate the cauliflower into smaller florets and cook them, covered, in salted water until al dente. Drain off the water and mash the cauliflower lightly with the milk with a potato masher.

Mix in eggs, cream, and salt and pepper. Spread a layer of cauliflower mash at the bottom of the buttered dish. Cover with salmon and sprinkle the top with chopped dill. Cover the fish with the remaining cauliflower mash.

Bake for 40 minutes, until the gratin has developed a nice crust. Serve with clarified butter.

Boiled Plaice with LCHF Spinach and Cheese Béchamel

A smooth and creamy spinach béchamel with cheese is a flavorful accompaniment to fish. Here, a piece of cooked plaice fillet gets some green company.

Makes 2–3 servings

7 oz. (200 g) whole spinach, parboiled
4-inch (10-cm) piece of leek
1 clove garlic
3 tablespoons butter
1/4 teaspoon nutmeg
salt and pepper
3/4 cup + 2 tablespoons (200 ml) heavy cream
1/3 cup + 1 1/2 tablespoons (100 ml) full-fat cheese, grated
5 1/4 oz. (150 g) plaice fillet per serving

Defrost spinach or use fresh leaves. Julienne the leek and mince the clove of garlic. Sauté them in the butter without letting them brown. Mix in the drained spinach.

Season with the spices and let simmer for 5 minutes, stirring occasionally. Add the cream, bring to a boil, and then let simmer for 20 to 30 minutes. Dilute with more cream if the béchamel becomes too thick. Stir in the cheese right before serving.

Cook the plaice in a pan with a lid in a little water, some lemon, and salt; this only takes a few minutes. Remove the fish with a slotted spoon, let it drain, and serve on top of the spinach béchamel.

Herring Roulades *with Mustard and Leek*

Herring is rich in good-for-you omega-3 fatty acids, and that's a good enough reason to eat herring often. It also happens to be delicious and reasonably priced.

Vary the fillings by using grated horseradish and diced apple, for example, or tomato purée and chives, or smoked caviar and dill.

Makes 2–3 servings

8–10 small herring fillets
salt and pepper
1 tablespoon unsweetened mustard + 1 teaspoon for the cream
2 tablespoons leek, finely julienned
2/3 cup (150 ml) heavy cream
1 teaspoon apple cider vinegar

Preheat the oven to 400°F (200°C). Grease an ovenproof dish. Cut off the dorsal fin on the herring fillets; season lightly with salt and freshly ground pepper on both sides of the fish. Spread 1 tablespoon mustard and the leek evenly over the fillet.

Roll up the fillets, skin-side in, and place them closely in the prepared ovenproof pan.

Lightly whisk the cream, and season with 1 teaspoon mustard, cider vinegar, and salt and freshly ground pepper. Pour this over the herring roulades and bake for 20 minutes.

Serve the herring roulades with a side of grated carrots dressed with freshly squeezed lemon juice.

Meat on the Menu

Cheese Salad with Pumpkin Seeds

A salad of full-fat cheese and pumpkin seeds is easy to assemble, and makes a tasty side dish for grilled chicken. If you find rotisserie chicken to be too spicy, try the oven-fried chicken on p. 122. Leftover chicken can be frozen in individual portions for later use.

Makes 1–2 servings

1/3 cup + 1 1/2 tablespoons (100 ml) creamy full-fat cheese, diced
1/4 cup (50 ml) hothouse cucumber, diced
2 tablespoons pumpkin seeds
1 tablespoon olive oil, cold pressed
1 teaspoon apple cider vinegar
salt and pepper
2 cherry tomatoes
chicken, grilled or fried

Mix cheese, cucumber, and pumpkin seeds with oil and vinegar. Season lightly. Quarter the tomatoes and sprinkle over the salad. Place next to the chicken and garnish with lettuce leaves and Italian flat-leaf parsley.

The mustard sauce on p. 74 works well here.

Pork Chipolatas with Sauerkraut and Fried Egg

The practice of fermenting vegetables was especially common before the advent of the icebox. Fermented white cabbage is a tasty and healthy addition to your LCHF menu.

I buy canned fermented sauerkraut and freeze it in smaller portions. The cabbage is good heated with sausage, pork knuckle, and ham. Add some variety to the sauerkraut by mixing in some caraway seed, juniper berries, apple, leek, or some crisply fried julienned bacon.

Makes 1 serving

4–5 pork chipolatas
2 tablespoons butter
1 egg
1/3 cup + 1 1/2 tablespoons (100 ml) sauerkraut
1/4 teaspoon caraway seeds

Snip a cross at the end of each chipolata. Fry the sausages in butter until golden brown and crispy, together with the egg. In a separate saucepan, heat the sauerkraut gently and season with caraway seeds. Garnish with a sprig of parsley.

Oven-Fried Chicken

Not long ago, oven-fried chicken was a Sunday dinner favorite. Even today it's still very festive, especially when served with smooth cream gravy, some jelly, and pickled cucumber. The difference between yesterday's chicken dinner and today's, however, is that I have made the jelly and the cucumber far less sweet. The fried, marble-sized potatoes have been swapped out for vegetables. The cream gravy contains no flour. Instead, we let it simmer to make the sauce thicken naturally without a roux.

Makes 4–6 servings

1 chicken, 2 ½–3 lbs. (1.2–1.4 kg)
3 tablespoons melted butter, for brushing
1 tablespoon salt
1/2 teaspoon pepper
1 teaspoon paprika
1 tablespoon dried parsley
1 tablespoon dried French tarragon

Preheat the oven to 435°F (225°C). Rinse the chicken inside and out with cold water, and leave the chicken propped upright in a colander to let it drain thoroughly.

Season the chicken cavity with salt and pepper, and bind the legs and wings against the body with twine. Brush the bird liberally with melted butter.

Mix the salt, pepper, paprika, dried parsley, and tarragon in a small bowl. Rub the spice mix thoroughly all over the chicken. Place the chicken, breast side down, on a rack in an ovenproof dish.

Bake in the oven for 20 minutes. Remove the chicken, turn the breast side up, and brush with more melted butter. Return the chicken to the oven, and lower the heat to 350°F (175°C). Let

the chicken bake another hour, brushing the chicken with butter from time to time. Check the chicken for doneness by pushing a sharp knife up to the breast bone. The chicken is ready if the meat juices run clear. If the meat juice is pink, however, leave the chicken in the oven a little while longer. If you use a meat thermometer, it should read 185°F (85°C) at the bone when the chicken is cooked.

Remove the chicken from the oven; cover it with foil and let it rest a little while before cutting into it. Deglaze the rest of the pan gravy with 1/3 cup + 1 1/2 tablespoons (100 ml) of water. Pass the gravy through a sieve and add the liquid to the cream sauce, or save it for future use in soup.

CREAMY GRAVY

2 cups (500 ml) heavy cream
1 1/4 cups (300 ml) whole milk
2 organic meat stock cubes
2 tablespoons black currant jelly (see recipe on p. 82)
2 teaspoons soy sauce (gluten-free)
1/4 teaspoon salt
1/4 teaspoon pepper

Mix all the ingredients in a thick-bottomed saucepan. Bring to a boil, stirring continuously. Let simmer over low heat for 45 minutes, stirring occasionally. Taste and adjust seasoning as needed. If the sauce thickens too much, dilute it with some milk. If it's too thin, turn up the heat and let it cook for a little while longer. Serve the sauce with the chicken.

Cooked cauliflower, carrots, and sugar snap peas are good side dishes. The recipe for pickled cucumbers is on p. 78.

Brined Pork Knuckle *with*

Mashed Rutabagas

Pork knuckle is another one of our homey classics. Cook it for a long time to ensure that the meat is fork tender and delicious.

The classic complement—rutabaga and potato mash—works in our LCHF kitchen too, because my refigured mash contains no potatoes.

Makes 3–4 servings

2.2 lbs. (1 kg) brined pork knuckle, bone-in
water to cover the pork knuckle
1 yellow onion
4-inch (10-cm) piece of leek
10 white peppercorns
6 allspice berries
2 bay leaves
1 small rutabaga
1 large carrot
1 large parsnip

MASHED RUTABAGA

cooked root vegetables from the pork knuckle
2 tablespoons butter
cooking liquid from pork knuckle
salt and pepper

Rinse the pork knuckle in cold water. Place it in a pot, cover it with cold water, and bring it to a boil. Skim off the foam thoroughly.

Cut the onion and leek into chunks, and add them to the meat in the pot along with the peppercorns, allspice berries, and bay leaves. Bring back to boil, and let simmer on low heat, covered, for 2–3 hours.

Cut the root vegetables in large pieces and add them at the last hour of cooking. When they're cooked, remove them with a slotted spoon; place them in another pot and mash with a potato masher or press.

Mix in some butter and cooking liquid. Adjust for seasoning with freshly ground pepper, perhaps some salt, and reheat carefully.

Sesame Crusted Pork Schnitzel

with Parmesan Butter and Cauliflower Mash

A schnitzel, with its double coating of wheat flour, egg, and breadcrumbs, is on many people's list of favorite foods. This revamped recipe contains far fewer carbohydrates, because I use sesame seeds instead of breadcrumbs. The result is nevertheless scrumptious.

Makes 1 serving

4 1/2 oz. (125 g) slice of pork, uncooked
salt and pepper
1 teaspoon psyllium husk powder
1 egg white
1/4 cup (50 ml) sesame seeds
3 tablespoons butter

Place the meat in a plastic bag and pound it to flatten it into the shape of a schnitzel; for this task I like to use a heavy saucepan. Season both sides of the schnitzel with salt and pepper, and then sprinkle the psyllium husk powder on both sides. The easiest way to do this is to use a salt shaker for the husk.

With a fork, whisk the egg white in a deep plate. Place the sesame seeds in a layer on a flat plate. Dip the meat in the egg white, making sure it covers both sides entirely. Hold the meat over the plate to drain.

Dip the schnitzel in the sesame seeds, and press on them to make them stick to the surface. Leave the schnitzel on the plate for a while to make the coating dry on the meat.

Lightly brown the butter in a frying pan and add the schnitzel to the pan. Turn down the heat a little, and fry for a few minutes on each side. Sesame seeds burn fairly easily, so keep an eye on the heat so the pan doesn't get too hot.

Serve the schnitzel with cauliflower mash (recipe follows) and Parmesan butter (see recipe on p. 77).

CAULIFLOWER MASH

3/4 cup + 2 tablespoons (200 ml) cauliflower florets
2 tablespoons heavy cream
1 tablespoon butter
salt and pepper

Cook the cauliflower in lightly salted water until soft. Drain the cauliflower thoroughly and press hard with a potato masher.

Carefully heat the mash with the cream and butter; season with salt and freshly ground pepper.

Meatloaf *with Mushrooms in LCHF Béchamel*

Meatloaf can be enjoyed both warm and cold. LCHF mushroom béchamel or cream sauce make equally good toppings when the meatloaf is served warm. I cut the cold loaf into thin slices and eat them as a sandwich with cheese and butter—a nutritious meal to eat along with a cup of coffee.

Makes 5–6 servings

1 yellow onion
1 3/4 oz. (50 g) hot-smoked pork belly
3 tablespoons butter, for frying and brushing
8 3/4 oz. (250 g) ground meat, from beef or game
8 3/4 oz. (250 g) ground pork
2 large eggs
2 tablespoons psyllium husk powder
1 teaspoon salt
1/4 teaspoon pepper
1/4 teaspoon nutmeg
1/8 teaspoon ground allspice
1/3 cup + 1 1/2 tablespoons (100 ml) heavy cream

Preheat the oven to 350°F (175°C). Butter an ovenproof dish. Chop the onion finely, and dice the pork belly. Fry in butter for a few minutes without letting the onion brown too much. Let it cool a little.

Meanwhile, mix the ground meats with eggs, psyllium husk powder, and spices. Add onion and pork belly, and mix thoroughly without working it too hard. Let stand 30 minutes, and then shape into a loaf.

Place the loaf in the prepared ovenproof dish and brush it with melted butter. Bake at 350°F (175°C) for 45 minutes.

The meatloaf will feel firm to the touch when it's ready. Test for doneness by sticking it with a toothpick to see if meat juices run clear. A meat thermometer should read 158°F (70°C) when the meatloaf is cooked.

Let the meatloaf rest for a while in the dish before slicing it. Serve with LCHF mushroom béchamel, cooked sugar snap peas, and tomatoes.

MUSHROOMS IN LCHF BÉCHAMEL

Makes 4 servings

8–10 mushrooms, sliced thinly
1/4 cup (50 ml) leek, julienned
3 tablespoons butter
salt
2 teaspoons pink peppercorns, crushed
1/2 organic vegetable stock cube
1 teaspoon dried French tarragon
1/4 cup (50 ml) white wine, or 1 teaspoon apple cider vinegar
3/4 cup + 2 tablespoons (200 ml) heavy cream

Brown the mushrooms and leek in the butter to give them some color; season with salt and crushed pink peppercorns. Crumble in stock cube and tarragon.

Pour in wine or vinegar, and let liquid cook into the mushrooms. Add cream, bring to a boil, and simmer for 20 minutes. Stir occasionally. Dilute with extra cream if too thick.

Hamburger *with Creamed White Cabbage*

When I mix ground meat, I like to make enough to prepare hamburgers and meatballs for many meals. I freeze them once they've been cooked, and reheat them carefully as needed.

In times past, white cabbage in béchamel was a common sight on any diabetic's menu; potatoes were banned. White cabbage doesn't spike blood sugar as potatoes do, so it's extremely compatible with today's focus on low-carbohydrate nutrition.

Makes 3–4 servings

1 yellow onion, finely chopped
2 tablespoons butter
8 3/4 oz. (250 g) ground meat, from game or beef
8 3/4 oz. (250 g) ground pork
1 large egg
1/4 cup (50 ml) heavy cream
1 1/2–2 teaspoons salt
1/8 teaspoon white pepper
1/8 teaspoon ground allspice
1/8 teaspoon ground ginger
1 teaspoon psyllium husk powder
3 tablespoons butter
salt and pepper

Cook onion in 2 tablespoons butter until soft. Let cool. Mix all ground meat with egg and cream. Stir in spices, psyllium husk powder, and sautéed onion. Mix thoroughly and let rest in the refrigerator for half an hour.

Shape small hamburgers or meatballs out of the meat. To keep meat from sticking, wet your hands in cold water. Set the hamburgers/meatballs down on a water-rinsed, wet cutting board.

Brown hamburgers in 3 tablespoons butter for a few minutes on each side. Start on high heat, then lower it after turning the hamburger, and let cook for another 5 minutes; season lightly with salt and pepper.

If making meatballs, don't crowd too many in the frying pan at a time, as the pan will cool down and they won't brown. Shake the pan while cooking so they keep their shape. Meatballs cook slightly quicker.

WHITE CABBAGE IN LCHF BÉCHAMEL

2 cups (500 ml) white cabbage
2 tablespoons butter
salt and pepper
1/2 organic vegetable stock cube
3/4 cup + 2 tablespoons (200 ml) heavy cream

Cut the cabbage into smaller pieces, or julienne coarsely. Fry in butter for 10 minutes over medium heat, without letting it brown. Season and crumble in the stock cube. Add cream and bring to a boil.

Let the béchamel simmer for at least 30 minutes, stirring occasionally. Test to see if the cabbage is soft; if not, cook it a little longer. If the béchamel thickens too much, dilute it with more cream or whole milk.

Carrots in LCHF Béchamel

with Chives

Cooking raw carrots in béchamel makes them retain their wonderful flavor. All their nutrition is preserved, too, instead of being lost to the cooking water.

Vary the seasonings to taste by adding dill, parsley, thyme, or horseradish.

Makes 2–3 servings

2 carrots
2 tablespoons butter
salt and pepper
1/3 cup + 1 1/2 tablespoons (100 ml) heavy cream
1/4 cup (50 ml) whole milk
2 tablespoons chives, chopped

Dice the carrots finely; sauté them in the butter for 5 minutes without letting them brown. Season with salt and freshly ground pepper, and pour in cream and milk. Bring to a boil, and let simmer over low heat for about 30 minutes, until the carrots are soft and the sauce is creamy.

Dilute the béchamel with some milk if it looks too thick. Stir in the chives at the end. Serve with fried smoked bologna and slices of pork belly.

Onion Sauce for Fried Pork Belly

Fried pork belly with onion sauce is a dish I often served my lunch guests throughout my restaurant career.

Nowadays I make the sauce without any flour at all, and the side dish is broccoli, not potatoes. The sauce tastes so much better and its carbohydrate count is far lower than before.

Makes 2–3 servings

1 yellow onion
2 tablespoons butter
salt and pepper
2/3 cup (150 ml) heavy cream

Slice the onion thin; sauté it in the butter for 5 minutes without letting it brown. Season it lightly with salt and freshly ground pepper.

Pour in the cream. Bring to boil, lower the heat and let it simmer for 15–20 minutes, until the onion is soft. Dilute with cream if the sauce is too thick.

Serve the sauce with crisp fried pork belly slices, and cooked broccoli or cauliflower.

Fricassée of Ham *with Lemon and Rosemary*

A mild fricassée (i.e., meat stew) is very tasty, and especially fitting for when you feel a little under the weather. Well-cooked meat in a gravy, thickened with egg yolk and cream, provides all the nourishment you'll need for a speedy recovery.

Makes 2 servings

8 3/4 oz. (250 g) boneless ham steak
3/4 cup + 2 tablespoons (200 ml) water
1 organic vegetable stock cube
juice of 1/2 lemon
1 bay leaf
5 white peppercorns
salt
2/3 cup (150 ml) heavy cream
1 tablespoon rosemary, finely chopped
2 egg yolks + 1/4 cup (50 ml) heavy cream

Dice the ham into 3/4-inch (2-cm) cubes. Bring the water and ham to a boil. Skim off the foam well. Add the stock cube, lemon juice, bay leaf, and peppercorns. Taste and add salt if needed.

Let simmer, covered, on low heat for about 1 hour, until the meat is fork tender. Move the pieces of meat to another saucepan and pass the cooking liquid through a sieve over the meat.

Bring to a boil with 2/3 cup (150 ml) cream and the finely chopped rosemary. Let simmer for 20 minutes. Mix in the rest of the cream with the egg yolks and add to the saucepan with the meat.

Let the fricassée simmer for only a short time, stirring constantly, as the yolks make the sauce thicken quickly. If it gets too hot the yolks might turn lumpy, so keep a close eye on the heat.

Serve with cooked cauliflower and carrots.

Calf Liver à l'Anglaise with Brown Butter

This dish was very popular restaurant fare at the beginning of the 1980s, when I started my career as chef. With the mild taste of liver and the piquant touch of smoked pork belly, beets, and capers, it isn't hard to understand why our diners loved it.

Makes 2–3 servings

8 ¾-oz. (250-g) piece of calf's liver
2 tablespoons butter
1/4 cup (50 ml) smoked pork belly, diced
2 tablespoons capers
1/4 cup (50 ml) pickled beet, diced
1 oz. (25 g) butter, for frying

Slice the liver. Melt and brown 2 tablespoons butter, to be served with the dish.

Fry the diced pork belly in its own fat in a small saucepan. Let it simmer until it is nicely browned. Stir in the capers and beets, and let it all heat through.

In a hot frying pan, sauté the liver in butter for a few minutes on each side. Remove from the flame and let the liver cook a little in the remaining heat of the pan.

Serve with a spoonful of the beet/caper mix, and dress with a spoonful of brown butter on top. Fried carrot and parsnip sticks make good complements.

Chicken Breast with Curry Sauce

Chicken in curry sauce was once served regularly in many homes. This version is made more up to date by using chicken breast, which is simpler to prep and also quicker to cook. The curry sauce is very mild, yet flavorful due to the inclusion of diced apple.

Makes 2–3 servings

2 chicken breasts
1/2 teaspoon salt
2 tablespoons leek, sliced
1 carrot, sliced
1 bay leaf
5 white peppercorns
a few sprigs of parsley
1 organic chicken stock cube

Barely cover the chicken breasts with salted water and bring to a boil. Skim the foam off. Add the vegetables, bay leaf, peppercorns, sprigs of parsley, and stock cube. Bring to a boil and let simmer, covered, over low heat for 20 minutes.

Remove the chicken breasts from the pan and cook the remaining cooking liquid to reduce it by half. Pass the stock through a sieve, and use 1/3 cup + 1 1/2 tablespoons (100 ml) of it for the sauce. The leftover can be saved to make a great base for soup.

CURRY SAUCE

1 small yellow onion
1 apple
3 tablespoons butter
1 tablespoon curry powder
1/3 cup + 1 1/2 tablespoons (100 ml) chicken stock
3/4 cup + 2 tablespoons (200 ml) heavy cream
salt and pepper

Finely chop the onion and dice the apple. Brown them lightly in butter, and let them cook together with the curry powder for a few minutes.

Add the stock and cream. Let the sauce cook to reduce for 30 minutes over low heat. Season with a little salt and pepper. Pass the liquid through a sieve to remove the bits of onion and apple.

Heat the chicken in the rest of the stock, while covered. To serve, cut the chicken diagonally and set the slices on a layer of curry sauce. Garnish with diced red bell pepper and a sprig of parsley or other green fresh herb. Serve with cauliflower rice.

CAULIFLOWER RICE

Shred cauliflower—about 3 1/3–6 3/4 fl. oz. (100–200 ml) per person—on a box grater's coarse grating surface. Cook it in lightly salted water for a few minutes; taste to see if it is ready. Drain the water and stir in a tablespoon of butter, some salt, and freshly ground pepper. Reheat carefully in the saucepan.

Carrot Gratin

Carrot gratin has had a place on almost every Christmas table of my 50-year-old life. My Finnish-Swedish mother brought this tradition over from her own childhood.

I've changed the gratin to make it more LCHF friendly. This recipe uses cottage cheese instead of rice, contains no wheat flour, and is less sweet overall.

2 large carrots
1/3 cup + 1 1/2 tablespoons (100 ml) cottage cheese
2 large eggs
1/3 cup + 1 1/2 tablespoons (100 ml) heavy cream
sweetener equivalent to 1–2 tablespoons honey
1/4 teaspoon nutmeg
salt and pepper

Preheat the oven to 350°F (175°C). Butter an oven-proof dish.

Cut the carrots into smaller pieces and cook them until soft in slightly salted water. Press the carrots in a potato ricer or with a potato masher, and mix in cottage cheese, eggs, and cream. Blend thoroughly until the mix is smooth.

Season with sweetener, nutmeg, salt, and freshly ground pepper. Pour the mixture into the prepared dish.

Bake for 40 minutes. Serve with clarified butter. Cooked ham, pork chipolatas, and meatballs are all delicious served with this carrot gratin.

Desserts

Chocolate Mousse

Dark chocolate mousse is a classic ending to a good dinner. It was very popular among our restaurant guests.

For anybody following LCHF, this mousse, made with egg yolk and heavy cream, is the perfect dessert.

Makes 4–6 servings

1 3/4 oz. (50 g) dark chocolate with 70–90 percent cacao
1 egg yolk
a few grains of salt
2/3 cup (150 ml) heavy cream

Melt the chocolate in a bowl set over a small saucepan of boiling water. Make sure no steam reaches the chocolate, or else it will turn grainy. Remove the bowl from the heat and let the chocolate cool a little.

Stir the yolk into the melted chocolate and mix thoroughly for a smooth mixture. Add a small pinch of salt to enhance the flavor of the chocolate.

Whip the cream—not too firm—until soft peaks form. Stir a spoonful of cream into the chocolate mix and blend thoroughly. Carefully fold the chocolate mixture into the remaining cream.

Fill small cups or glasses with the mousse, and store in the refrigerator for a few hours to set. Serve with piped whipped cream and garnish with a fresh raspberry and lemon balm.

This mousse will keep up for up to a week in the refrigerator if it is kept in a plastic container. It can be attractively presented with a spoon dipped in warm water, or piped into a glass with a piping bag.

Fried Apples *with Crème Chantilly*

Our local apples are so good that I occasionally enjoy eating one for dessert. And whipped cream with real vanilla powder makes a delicious replacement for the sugar-filled custard that usually goes with apples. Vanilla powder is the ground vanilla bean that you find in small jars in the baking aisle of well-stocked grocery stores. Vanilla powder is totally devoid of carbohydrates and chemical aftertaste, both of which are common in ordinary vanilla sugar.

Fruit tends to quickly spike blood sugar levels, but the response is not as dramatic if the fruit is eaten for dessert after a meal. Keep this in mind, especially if you are diabetic.

Makes 2–3 servings

1 apple
2 tablespoons butter
sweetener equivalent to 1/2 teaspoon honey
1/4 teaspoon cinnamon
1/3 cup + 1 1/2 tablespoons (100 ml) heavy cream
a pinch of real vanilla powder

Cut the apple into small cubes. Brown the butter lightly and fry the apple cubes for a few minutes together with the sweetener and cinnamon. Let the mixture cool, and serve it in small glasses with a dollop of vanilla whipped cream on top.

Cottage Cheesecake

An old-fashioned cheesecake gets a makeover when we use cottage cheese instead of wheat flour and rennet. While taste and texture are almost identical to the original, the carbohydrate content is dramatically reduced in this updated version.

Makes 2–3 servings

2 large eggs
1/3 cup + 1 1/2 tablespoons (100 ml) cottage cheese
1/4 cup (50 ml) almonds, chopped
2 bitter almonds, finely chopped (or a few drops of almond extract)
1/4 cup (50 ml) heavy cream
1/8 teaspoon real vanilla powder
a pinch of salt

Preheat oven to 350°F (175°C). Butter a small ovenproof dish.

Beat together eggs, cottage cheese, chopped almonds (and extract, if using), and cream. Flavor the batter with vanilla powder and a tiny bit of salt, and pour it into the prepared dish.

Bake for 40 minutes. Serve lukewarm, together with strawberry compote and lightly whipped cream.

STRAWBERRY COMPOTE

3/4 cup + 2 tablespoons (200 ml) strawberries, sliced
1 tablespoon lemon juice
sweetener equivalent to 1/2 tablespoon honey

In a saucepan, bring strawberries and lemon juice to a boil. Reduce the heat and let simmer for a few minutes, stirring occasionally. Remove the pan from the heat and let it cool a little, and add the sweetener. Pour the compote into a jar and store it in the refrigerator. It will stay fresh for up to 4–5 days.

Lingonberry Ice Cream

You don't need an ice cream maker to make my easy, homemade lingonberry ice cream; this recipe has far less sweetener than store-bought varieties.

The tart lingonberries are very good matched with smooth vanilla. This ice cream is of course delicious made with other berries, too.

Makes 8–10 servings

3 egg yolks
sweetener equivalent to 2 tablespoons honey
1 1/4 cups (300 ml) heavy cream
1/8 teaspoon real vanilla powder
3/4 cup + 2 tablespoons (200 ml) preserved uncooked lingonberries (or cranberries) (see recipe on p. 81)

Beat the yolks and sweetener until thick and full. Whip the cream with the vanilla powder. Stir the lingonberries into the egg yolk mixture, and then carefully fold in the whipped cream.

Pour the mix into small cups or one larger dish, and set in the freezer until the next day. Before serving, put the ice cream in the refrigerator for half an hour to make the texture creamy.

If using a larger dish, unmold it by first dipping the container quickly into warm water. Stick a fork into the middle of the ice cream and turn the fork slightly – this should make the ice cream easy to unmold.

Cottage Cheese à la Malta with Queen Sauce

Cottage cheese is not only very tasty and useful but also contains lots of protein. Here, it replaces white rice in our version of Rice à la Malta. Add a few chopped walnuts and you'll increase its nutritional value further still.

Instead of serving it with a sugary berry sauce, drizzle our fresh and delectable unsweetened berry sauce on it.

Makes 2–3 servings

1/3 cup + 1 1/2 tablespoons (100 ml) cottage cheese
1/4 cup (50 ml) walnuts, chopped
1/8 teaspoon real vanilla powder
1/3 cup + 1 1/2 tablespoons (100 ml) heavy cream

Mash the cottage cheese with a fork into smaller curds. Mix it with chopped walnuts and vanilla powder. Whip the cream until soft peaks form. Carefully fold the cottage cheese mixture into the whipped cream. Set in the refrigerator for a spell to let the vanilla flavor develop fully.

QUEEN SAUCE

1/3 cup + 1 1/2 tablespoons (100 ml) raspberries
1/3 cup + 1 1/2 tablespoons (100 ml) bilberries (blueberries)
2 teaspoons lemon juice

Mix the berries and lemon juice in a saucepan. Bring the mixture to a boil and let simmer for a few minutes. Pass the cooked berries through a fine mesh sieve by pressing on them with a ladle or spoon. Pour the juice into a jar and let it cool. Store in the refrigerator.

This sauce doesn't keep as well as one full of sugar. If you can't use it up within a few days, store the jar in the freezer.

Jellied Raspberry Mousse

Years ago, different versions of jellied mousse were very popular for dessert. It's not often you'll see a recipe for one of those nowadays, however, so I've included a favorite of mine featuring the tang of wild raspberries. I've also used this mousse in layer cakes many times throughout my restaurant career.

To put together such a layer cake, use the recipe for cardamom cake on p. 173. Leave out the cardamom and you'll have the perfect base for a layer cake. The mousse keeps for several days in the refrigerator.

Makes 6–8 servings

2 sheets of gelatin (or about 1 1/2 teaspoons powdered gelatin)
1 1/4 cups (300 ml) raspberries
2 large eggs
sweetener equivalent to 1 tablespoon honey
3/4 cup + 2 tablespoons (200 ml) heavy cream
1 teaspoon lemon juice

Soak the sheets of gelatin (if using powder, follow instructions on the packet) in cold water for 10 minutes. Bring the raspberries to a quick boil, and then pass them through a fine mesh sieve by pressing down on them with a ladle or spoon. Let the purée cool.

Use two bowls to separate the eggs into yolks and whites. Mix the yolks with the sweetener and beat with a handheld electric mixer until the mixture is thick and voluminous. Whip the cream until soft peaks form.

If using the same beaters, wash them thoroughly before whipping the egg whites – if there's any fat left on them, the whites will not whip up to desired stiffness.

Press the gelatin leaves to remove excess water (for powdered gelatin, follow packet instructions) and melt them together with the lemon juice over low heat.

Stir the melted gelatin into the raspberry purée, and then add the mixture to the egg yolk. Fold in the whipped cream, and finally, carefully fold in the egg whites.

Fill small cups or delicate glasses with the mousse, or fill a larger dish with the entire amount of mousse. Place the mousse in the refrigerator for a few hours to allow it to set. Serve with whipped cream.

Whipped Cream Pastry

Just because we shy away from unnecessary carbohydrates doesn't mean that we can't indulge in a delicious treat at special occasions or a party.

What could taste better than a whipped cream pastry with your coffee?

Makes 2 servings

1/3 cup + 1 1/2 tablespoons (100 ml) heavy cream
2 raspberry "grottos" (see recipe on p. 162)
fresh or frozen raspberries
lemon balm

Whip the heavy cream until stiff peaks form, and put it in a piping bag. Spread a layer of whipped cream onto the grottos and pipe with the whipped cream up the side.

Garnish with raspberries and a sprig of lemon balm.

A Small Cookie with Your Coffee

Raspberry Grottos

The proffered cookie plate at a coffee klatch is enriched by the inclusion of these raspberry-filled almond cookies. I add some more mashed raspberries to them just before serving to make them even more irresistible. You can also use any other types of berries if you like.

These cookies can be turned into delicious whipped cream pastries by following the recipe on p. 158.

Makes 12 grottos

1 3/4 oz. butter, cold
3/4 cup + 2 tablespoons (200 ml) almond meal
1 large egg
sweetener equivalent to 1 tablespoon honey
1 tablespoon psyllium husk powder
1 teaspoon baking powder
2/3 cup (150 ml) mashed raspberries

Preheat the oven to 400°F (200°C).

Cut the butter into small chunks and mix it with the ground almonds. (It's easiest to do this with a hand-held electric mixer.) Stir in egg and sweetener. Mix psyllium husk powder and baking powder and add them to the batter.

Let the batter rise for 5 minutes in the refrigerator, then roll it out, turn into small balls, and set them in cupcake papers. Make a deep well at the center of each cookie and fill it with a teaspoon of raspberry mash.

Bake for 8–10 minutes. Leave to cool on a wire rack. These cookies freeze well. Set them in a warm oven right before serving and they'll taste as if they're freshly baked.

No-Bake Chocolate-Dipped Macaroons

These coconut cookies taste like fine candy; decorated with white chocolate, they deserve a place on your coffee table.

Makes 25 cookies

1 3/4 oz. (50 g) butter, room temperature
sweetener equivalent to 1 tablespoon of honey
1/4 teaspoon real vanilla powder
1/3 cup + 1 1/2 tablespoons (100 ml) heavy cream
7 oz. (200 g) shredded coconut, unsweetened
3 1/2 oz. (100 g) dark chocolate with 70–90 percent cacao
4 squares sugar-free white chocolate

With a handheld electric mixer, beat butter, sweetener, and vanilla powder until light and airy.

Add the cream while continuing to mix. Add the shredded coconut, and set the dough in the refrigerator for 30 minutes to firm up.

Roll into tablespoon-sized balls, place them on a platter, and flatten them a little. Chill the balls for 10 minutes before dipping them in chocolate.

In separate bowls, melt the dark and white chocolate over boiling water. Make sure no steam reaches the chocolate, or else it will turn grainy. Dip the

cookies in the dark chocolate, and then set them on a wire rack.

You'll have an easier time dipping the cookies by using two forks: Spear the cookie on one fork, and dunk it quickly in the chocolate. Tap the fork lightly against the edge of the bowl to remove the excess chocolate drips, and push the cookie on to the wire rack with the second fork.

Make a small piping bag out of wax paper, or use a plastic bag with a small hole cut into one corner. Fill the bag with white chocolate and pipe a spiral starting along the outer edge of each cookie. With a toothpick, draw lines toward the middle of the cookie to make a flower motif.

Place the cookies in the refrigerator to chill. Store them for up to a week in a tin in the refrigerator, or freeze and defrost them as needed (in the refrigerator). Bring the cookies to room temperature for a bit before serving them to let their flavor develop fully.

Spice Cookies

Small cookies full of richly spiced flavors are an integral part of the holiday cookie platter. In fact, I promise you they're delicious with a cup of coffee any time of the year.

Makes 20 cookies

3 1/2 oz. (100 g) butter
1/2 teaspoon ground cardamom
1/2 teaspoon cinnamon
1/2 teaspoon ground cloves
1/4 teaspoon ground ginger
sweetener equivalent to 2 tablespoons of honey
3/4 cup + 2 tablespoons (200 ml) almond meal
2 tablespoons psyllium husk powder
1 large egg

Preheat oven to 400°F (200°C). Line a cookie sheet with parchment paper.

Put butter, spices, and sweetener in a saucepan. Bring to a quick boil, stir thoroughly, and then let it cool.

Mix the almond meal with the psyllium husk powder in a bowl. Add the spice mixture and the egg. Mix thoroughly. Leave in the refrigerator to rise for 5 minutes.

Roll out tablespoon-sized balls from the batter and flatten them slightly. Place the balls on the prepared cookie sheet and make a crisscross pattern on the cookies with a fork.

Bake for 10–12 minutes. Let cool on a wire rack.

Freeze and defrost the cookies as needed. Once they're defrosted, set them in a warm oven for a few short minutes to crisp up their surface.

Scalloped Almond Shells

My mother often baked scalloped almond shells when the mood was festive. They were almost always stuffed with "forest gold," as we had plenty of cloudberries from the bogs in the mountainous areas near our farm.

Makes 10–12 cookies

1 3/4 oz. (50 g) butter, cold + 1 tablespoon melted butter for the shell tins
3/4 cup + 2 tablespoons (200 ml) almond meal
2 teaspoons psyllium husk powder
sweetener equivalent to 1 tablespoon honey
1 large egg

Preheat the oven to 400°F (200°C). Grease the scalloped cookie tins and place on a baking sheet.

Cut the butter into small pieces. Mix it quickly with the almond meal, psyllium husk powder, and sweetener. The easiest way to do this is to use a handheld electric mixer. Finally, mix in the egg.

Let the dough rise for 10 minutes in the refrigerator. Fashion tablespoon-sized balls and press them into the scalloped metal tins.

Bake for 8–10 minutes.

These almond shells are not going to be as tender as the ones baked of wheat flour, which makes them easier to remove from their tins once they've cooled. Freeze the cookies that you're not planning on using immediately. Once they're defrosted, set them in a warm oven for a few short minutes to crisp up their surface. Serve with whipped cream and berries.

Chocolate-Drizzled Nut Cookies

A cookie can be delicious and nutritious at the same time. Here you'll find a happy mix of nuts, seeds, and dark chocolate alongside cinnamon and cardamom.

Makes 20 cookies

2 large eggs
sweetener equivalent to 2 tablespoons of honey
3/4 cup + 2 tablespoons (200 ml) hazelnuts, finely chopped
1/3 cup + 1 1/2 tablespoons (100 ml) sunflower seeds, roasted
1 tablespoon psyllium husk powder
2 teaspoons cinnamon
1 teaspoon ground cardamom
1/4 teaspoon real vanilla powder
a pinch of salt
1 3/4 oz. (50 g) butter, melted
4 squares of dark chocolate, melted
10 hazelnuts, halved

Preheat the oven to 400°F (200°C). Place cookie molds on a baking sheet.

With a handheld electric mixer, beat eggs and sweetener until light and airy. Mix all dry ingredients in another bowl; add to the egg batter. Finally, stir in the melted butter. Using two spoons, fill the small cookie molds with batter. Bake for 15 minutes. Let the cookies cool on a wire rack.

Garnish each cookie with some drizzled chocolate and top with half a hazelnut.

Cardamom Cake

I've tried different ways to bake a cardamom cake without wheat flour. This version, made with ground almonds, is the best; it's now a new favorite in my kitchen.

The cake makes an excellent base for a layer cake if you leave out the cardamom.

Makes 1 loaf

butter and sesame seeds for the cake pan
3 1/2 oz. (100 g) butter, room temperature
sweetener equivalent to 2 tablespoons of honey
3 large eggs
3/4 cup + 2 tablespoons (200 ml) almond meal
2 tablespoons psyllium husk powder
1 teaspoon baking powder
1/4 teaspoon real vanilla powder
2 teaspoons ground cardamom
1/3 cup + 1 1/2 tablespoons (100 ml) heavy cream

Preheat the oven to 350°F (175°C). Butter a loaf pan, and sprinkle evenly with sesame seeds. Bring the butter to room temperature for a while before starting.

With a handheld electric mixer, beat the butter and sweetener vigorously until it is light and fluffy. Add the eggs, one at a time, to the butter/sweetener mix. Beat hard between each addition to thoroughly mix the butter and eggs.

Mix the ground almonds with the psyllium husk powder, baking powder, vanilla powder, and cardamom. Stir it into the batter together with the cream.

Bake for about 30 minutes until the cake feels firm to the touch.

Remove the pan from the oven and let the cake rest for a while before unmolding it on a platter. Run a thin spatula all around the pan to loosen the cake, and lift the cake gently from the bottom before turning it upside down. Leave the cake to cool; it will be moist and tender if you let it cool covered by the cake pan.

Peppermint-Flavored Ice Chocolate

Ice chocolate candy fits perfectly within LCHF nutrition guidelines. Organic coconut fat contains many nutrients that have proven to be beneficial for cognitive health. So go ahead and enjoy the occasional piece of ice chocolate with your coffee, and not just at the holidays.

Makes 10–12 pieces

3 1/2 oz. (100 g) dark chocolate with 70–90 percent cacao
1 3/4 oz. (50 g) organic coconut fat
a few drops of food grade peppermint oil
a few salt grains
1 large egg yolk

Melt the chocolate and coconut fat over a water bath or in the microwave. Flavor it with a few drops of peppermint and a pinch of salt; the salt will enhance the chocolate flavor. Stir in the egg yolk and mix thoroughly.

Double up some paper candy molds and set them on a tray; I prefer to use paper molds instead of foil because it's more environmentally friendly.

With a teaspoon, fill the molds with the chocolate mix.

You can vary the flavors by adding, for example, chopped walnuts, grated orange peel, or ground cardamom.

Examples of Items in the LCHF Kitchen's Refrigerator and Freezer

Dairy:
• Butter, cheese, eggs, cottage cheese, heavy cream, crème fraîche, cultured and sour cream, full-fat yogurt, mayonnaise

Meat, Fish, and Deli:
• Beef: beef chuck, stew meat, sirloin, brisket, ground beef, calf's liver
• Pork: fresh ham, pork shoulder, cured pork belly, pork shank, ground pork, pork loin, chops, ribs
• Lamb: chops, stew meat, lamb roast
• Game: ground meat, steaks, roast, stew meat
• Poultry: fresh whole chicken, chicken breast, chicken thighs, grilled chicken, ground chicken, turkey breast
• Fish and Shellfish: salmon, Baltic herring, herring, cod, plaice, shrimp, crayfish, mussels, salty/sweet cured and smoked salmon, smoked Baltic herring, pickled herring, anchovies, Swedish smoked caviar spread, roe
• Deli: sausage with high percentage of meat, hot-smoked pork belly, bacon, calf's headcheese, smoked turkey, cooked and smoked ham, salami, bologna, smoked reindeer meat, liver paté, roast beef

Vegetables and Fruit:
• Broccoli, cauliflower, white cabbage, carrots, parsnips, rutabagas, onions, leeks, garlic, tomatoes, cucumbers, bell peppers, mushrooms, spinach, avocados, lemons, apples, pears

Fresh and Frozen Vegetables and Berries:
• Broccoli, cauliflower, green beans, asparagus, spinach, sugar snap peas, dill, parsley, chives, cloudberries, lingonberries, cranberries, bilberries (blueberries), raspberries, strawberries

Pantry Staples, Canned Goods, and Other Convenience Items:
• Almonds, walnuts, hazelnuts, pumpkin seeds, sunflower seeds, sesame seeds, flax seeds, shredded coconut (unsweetened), psyllium husk powder, almond flour, coconut fat (organic), real vanilla powder, sweeteners such as erythritol, stevia, Xylitol, dark chocolate with 70–90 percent cacao; all these items can be found in most well-stocked grocery stores or health-food stores.
• Mackerel and sardines in tomato sauce, water-packed tuna, anchovies
• Mustard (unsweetened), sauerkraut, asparagus, green beans, canned tomatoes, olives
• Olive oil (cold pressed)
• Apple cider vinegar, red or white wine vinegar
• White vinegar (or 12 percent acetic acid)
• Spices and Dried Herbs: salt, salt flakes, white pepper, black pepper, allspice, bay leaves, cloves, ginger, cinnamon, cardamom, anise, fennel, caraway seeds, rosemary, thyme, French tarragon, curry powder, organic stock cubes

Comments from Annika Dahlqvist, MD

Breakfast

The very best breakfast you can eat is egg-based. Eggs are the perfect food, as they contain all the nourishment a developing chicken requires, which is also all the nutrition we need, too. You can live on eggs alone, although it's preferable to have a side of vegetables. Eggs contain some cholesterol, but the body only absorbs what it needs; if it doesn't get enough of it from the food you ingest, it will manufacture the rest by itself. It's impossible to have high cholesterol by eating too many eggs.

Snacks

Cheese (with a pat of butter, perhaps), an egg, and a handful of nuts are examples of good-for-you snacks or light evening meal.

The Main Meal

The appropriate size of your daily intake of protein—whether derived from meat, fish, cheese, eggs, beans, or lentils—is equal to that of a closed fist, once or twice a day.

How much fat you need depends on each individual, so you'll have to determine that level yourself. The main guideline is don't take in more than you need for adequate nutrition and energy. If you eat 3 ½ oz. (100 g) of fat, 2 ¾ oz. (82 g) protein (keep in mind that meat only contains 20 percent protein), and ¾ oz. (20 g) carbohydrates a day, you'll have taken in approximately 1,300 calories. So if you feel that you need more calories, add some fat and possibly some carbohydrates, especially if your work involves manual labor.

To calculate energy value (calories), multiply grams of fat by 9, and grams of protein and carbohydrates by 4.

It's a good idea to pair meat and sauce with, for example, white or other cabbages, which are rich in vitamin C and minerals.

Beverages

Water is the best beverage, no doubt about it. For a refreshing taste, add some lemon or cucumber slices to your glass. Sugared drinks are full of empty, blood-glucose-elevating calories.

Let's use caution with alcohol; it's poisonous for the brain, liver, and pancreas. However, if you have no problem handling alcohol, a glass of wine now and then, once a week at most, will not do you any harm.

If you have questions about LCHF that you cannot find the solutions to in this book, feel free to go to my blog, and I will answer them there: www.doktordahlqvist.se

LCHF Menu for 3 Weeks

You don't have to vary your breakfast every day. Many are perfectly happy—indeed, some prefer—eating the same thing each morning. Coffee or tea is allowed, of course—why not drink it with a splash of full-fat cream?

Week 1   **BREAKFAST**

Monday	Canned mackerel in tomato sauce with cottage cheese
Tuesday	Bacon and eggs with half a fried tomato
Wednesday	Almond bread with cheese, slices of turkey, and bell pepper
Thursday	Canned sardines with cottage cheese and mayonnaise
Friday	Cold meatloaf with cheese and pickled cucumber
Saturday	Pickled herring with hard-boiled egg and cheese bread
Sunday	Yogurt (preferably 10–17 percent fat) with bilberries (blueberries) and LCHF muesli

Week 2

Monday	Galettes (Swedish pancakes) with bacon and uncooked lingonberry preserves
Tuesday	Scrambled eggs with slices of smoked reindeer roast
Wednesday	Buckwheat porridge with raspberries and a dollop of butter
Thursday	Bologna with fried egg on seed bread
Friday	Calf brawn (headcheese) with beet salad
Saturday	Hard-boiled egg with Swedish smoked caviar spread (or canned roe) and cottage cheese
Sunday	Soused herring with Swedish cultured cream (or sour cream) and chives

Week 3

Monday	Yogurt (preferably 10–17 percent fat) with cottage cheese, apple, and cinnamon
Tuesday	Canned tuna with hard-boiled egg and mayonnaise

Wednesday	Slices of cheese and ham on cheese bread with bell pepper
Thursday	Canned sardines in tomato sauce with cottage cheese and mayonnaise
Friday	Pork chipolatas with fried egg and tomato
Saturday	Waffles with smoked salmon in egg butter
Sunday	Bacon and scrambled eggs

Week 1   LUNCH

Monday	Fried pork belly rashers and bologna with carrots in LCHF béchamel
Tuesday	Salmon with cauliflower in LCHF dill béchamel
Wednesday	Fried pork belly and onion sauce
Thursday	Fish and shellfish soup
Friday	Boiled brook trout with dill mayonnaise and broccoli
Saturday	Curry chicken and cauliflower rice
Sunday	Ham fricassée with lemon and rosemary. Dessert: chocolate mousse

Week 2

Monday	Pork chipolata with fried egg and sauerkraut
Tuesday	Oven-fried Baltic herring with onion sauce and broccoli
Wednesday	Grilled chicken with cheese salad
Thursday	Spinach soup with half a hard-boiled egg
Friday	Cooked cod with egg sauce
Saturday	Calf liver à l'Anglaise with carrots and parsnip
Sunday	Pork schnitzel with cauliflower mash. Dessert: fried apples with Crème Chantilly

Week 3

| Monday | Brined pork knuckle with mashed rutabaga |

Tuesday	Hamburger with creamed white cabbage
Wednesday	Salmon gratin with clarified butter
Thursday	Old soupy stew with salt pork and white cabbage
Friday	Cooked plaice with spinach in LCHF béchamel
Saturday	Meatloaf with mushrooms in LCHF béchamel
Sunday	Chicken in cream sauce, pickled cucumber, and jelly. Dessert: Raspberry mousse

Week 1   DINNER/EVENING MEAL

Monday	Almond bread with liver paté and pickled cucumber
Tuesday	Yogurt (preferably 10–17 percent fat) with raspberries and muesli
Wednesday	Smoked salmon with scrambled eggs
Thursday	Hamburger with onion on seed bread
Friday	Cottage cheese à la Malta with queen sauce
Saturday	Salmon salad with hard-boiled egg and avocado
Sunday	Calf brawn (headcheese) with hard-boiled egg and beets

Week 2

Monday	Turkey roulades with pear cottage cheese
Tuesday	Meatball sandwich with beet salad
Wednesday	Smoked Baltic herring gratin with mushrooms
Thursday	Cheesecake with strawberry compote
Friday	Carrot gratin with cooked ham
Saturday	Waffles with cloudberries and heavy cream
Sunday	Salt and sweet cured salmon with mustard sauce and hard-boiled egg

Week 3

Monday	Cheese bread with smoked Baltic herring and scrambled eggs
Tuesday	Galettes (Swedish pancakes) with bilberries (blueberries) and crème fraîche
Wednesday	Smoked reindeer roast with horseradish cream
Thursday	Waffle with Swedish smoked caviar spread (or canned roe), chives, and Swedish cultured cream (or sour cream)
Friday	Buckwheat porridge with lingonberries (cranberries) and a dollop of butter
Saturday	Cooked asparagus with smoked caviar spread (or canned roe) butter
Sunday	Gentleman's Delight with cheese bread

Recipe Index